ENVIRONMENT AND MAN

VOLUME EIGHT

The Built Environment

ENVIRONMENT AND MAN: VOLUME EIGHT

Titles in this Series

Volume 1	*Energy Resources and the Environment*
Volume 2	*Food, Agriculture and the Environment*
Volume 3	*Health and the Environment*
Volume 4	*Reclamation*
Volume 5	*The Marine Environment*
Volume 6	*The Chemical Environment*
Volume 7	*Measuring and Monitoring the Environment*
Volume 8	*The Built Environment*

ENVIRONMENT AND MAN
VOLUME EIGHT

The Built Environment

General Editors

John Lenihan

O.B.E., M.Sc., Ph.D., C.Eng., F.I.E.E., F.Inst.P., F.R.S.E.

Director of the Department of Clinical Physics and Bio-Engineering, West of Scotland Health Boards, Professor of Clinical Physics, University of Glasgow.

and

William W Fletcher

B.Sc.,Ph.D., F.L.S., F.I.Biol.,F.R.S.E.

Professor of Biology and Past Dean of the School of Biological Sciences, University of Strathclyde.

1978

ACADEMIC PRESS · NEW YORK & SAN FRANCISCO

A Subsidiary of Harcourt Brace Jovanovich, Publishers

Blackie & Son Limited
Bishopbriggs
Glasgow G64 2NZ

450/452 Edgware Road
London W2 1EG

© 1978 Blackie & Son Ltd.
First published 1978

*All rights reserved.
No part of this publication may be reproduced,
stored in a retrieval system, or transmitted,
in any form or by any means,
electronic, mechanical, recording or otherwise,
without prior permission of the Publishers*

International Standard Book Number

0-12-443508-4

Library of Congress Catalog Card Number

~~75-37435~~ wrong #

Printed in Great Britain

Background to Authors

Environment and Man: Volume Eight

TONY J. CHANDLER, M.Sc., Ph.D., is Professor of Geography at the University of Manchester. He is a member of the Royal Commission on Environmental Pollution, the Clean Air Council, and the Natural Environment Research Council. He was formerly Secretary and subsequently Vice-President of the Royal Meteorological Society.

FIKRY N. MORCOS-ASÄAD, B.Arch., M.Arch., S.M. (M.I.T.), Ph.D., M.A.S.C.E., R.I.B.A., F.R.I.A.S., is Professor of Architecture in the Department of Architecture and Building Science, University of Strathclyde. He practised in Egypt, the United States and Britain, and was Consultant Architect to the American Embassy in Cairo. He is also a Commissioner of the Royal Fine Art Commission for Scotland, and is registered with the United Nations International Roster of Experts.

ROBERT WHITE, B.Sc., F.I.C.E., F.I.H.E., F.R.S.A., is Emeritus Professor of Civil and Transport Engineering of the University of Newcastle upon Tyne. He has been a member of the Road Research Laboratory Advisory Committee on Traffic Research, and a member of the Committee of the Universities' Transport Study Group.

JAMES McL. FRASER, B.Sc., C.Eng., F.I.C.E., is a partner in Babtie Shaw and Morton, Consulting Engineers, and titular head of their Public Health Engineering Division. He is a Consultant to the World Health Organization.

JAMES K. FEIBLEMAN is Bingham Professor of Humanities at the University of Louisville, Kentucky. A prolific writer and one of the most respected philosophers in the world today, he taught at Tulane University for 32 years and served for 19 years as Chairman of the School's Philosophy Department. Most recently he was Andrew W. Mellon Professor in Humanities at Tulane University. He is a past President of the Charles S. Peirce Society and the New Orleans Academy of Sciences.

Series Foreword

MAN IS A DISCOVERING ANIMAL—SCIENCE IN THE SEVENTEENTH CENTURY, scenery in the nineteenth and now the environment. In the heyday of Victorian technology—indeed until quite recently—the environment was seen as a boundless cornucopia, to be enjoyed, plundered and re-arranged for profit.

Today many thoughtful people see the environment as a limited resource, with conservation as the influence restraining consumption. Some go further, foretelling large-scale starvation and pollution unless we turn back the clock and adopt a simpler way of life.

Extreme views—whether exuberant or gloomy—are more easily propagated, but the middle way, based on reason rather than emotion, is a better guide for future action. This series of books presents an authoritative explanation and discussion of a wide range of problems related to the environment, at a level suitable for practitioners and students in science, engineering, medicine, administration and planning. For the increasing numbers of teachers and students involved in degree and diploma courses in environmental science the series should be particularly useful, and for members of the general public willing to make a modest intellectual effort, it will be found to present a thoroughly readable account of the problems underlying the interactions between man and his environment.

Preface

THE BIOLOGICAL CHARACTER OF THE HUMAN SPECIES HAS NOT CHANGED FOR 40,000 years. But since the end of the last Ice Age, 11,000 years ago, man has invented an environment which is almost entirely artificial, with the consequence that he now lives in a world virtually of his own devising.

Professor Tony Chandler offers a detailed analysis of one aspect of this man-modified environment. One fifth of us live in cities of more than 100,000 people, and by the end of the century 60% of the world's population will be living in urban environments.

The significance of these observations is that buildings, though designed primarily to provide a comfortable internal climate, have an appreciable influence on the external climate. Towns are heat islands; a city centre is generally about 1°C warmer than the surrounding rural areas by day and 2° warmer by night. City dwellers therefore have the modest advantages of smaller heating needs and a longer growing season for plants—but air pollution (mainly produced by motor vehicles, with a relatively small contribution from industry) leads to increased rainfall and a deterioration in the amount and quality of sunlight.

Buildings are among the most complicated artifacts produced by man and have had a major influence on the development and spread of civilization. Richly varied architectural achievements of temperate regions are not relevant to the needs of tropical countries, where the main problem facing the designer of a building is how to keep the occupants cool.

Rural communities have over the centuries devised ingenious solutions based on indigenous materials and technologies. But the increasing urbanization of the arid tropical lands presents the architect with new and formidable problems. Professor Fikry Morcos-Asäad, drawing on experience in both tropical and temperate regions, emphasizes that the optimum design of the built environment requires close study of the associated cultural and social environments, as well as of the more obvious aids and influences provided by present-day science and technology. After reviewing the impact of climate on human comfort and the thermal properties of buildings, he examines in detail the problems of securing satisfactory internal environments in countries characterized by various degrees of heat and humidity. His theoretical considerations are abundantly illustrated by practical applications and results.

Professor Robert White examines the total impact of transport on the environment. Analysing the movements of vehicles and people, he finds that the road patterns of most cities have not altered enough in response to social and industrial change and are now little related to modern needs. The major disturbances associated with road traffic—accidents, noise, pollution and visual intrusion—are all enhanced by the great concentration of vehicles in city centres, where most of them do not have any business to perform.

Possible control measures are of two kinds—construction and restriction. Accident rates (per million vehicle-miles) are highest in city centres and lowest on motorways. But how are vehicles to be kept out of city centres? Professor White considers several possibilities and concludes that there is no easy answer. Most people want to live at low density in near-rural surroundings, yet need access to the amenities of a large city. In this situation no public transport system can be economically viable.

The resolution of the urban transport problem remains an uneasy compromise among the conflicting requirements of personal convenience, economy and amenity.

Man (according to a recent World Health Organization publication) cannot live in dignity in the midst of his own waste. Mr. James Fraser, writing in appropriately forthright style, shows how technology can cope with the torrent of waste produced from the body, the home, the factory and the farm.

The emperor Vespasian enriched his treasury by a tax on the urine which was collected by the fullers and launderers of Rome and used for its detergent properties. When his son objected to such fiscal ingenuity, Vespasian held up a bag of gold and remarked: "Non redolet". To most people, however, body waste is offensive. Mr. Fraser describes the effective procedures now used to render human excreta harmless—and even useful.

Industrial wastes present a more difficult problem. The wide variety of material discharged from factories—some poisonous, some acid, some alkaline and some greedy for oxygen—can, if not promptly treated, do great damage to living systems. The hazards are growing, but are still being contained.

Looking to the future, Mr. Fraser suggests that the waste disposal problem would be eased by limiting the built-in obsolescence which now shortens the life of many products. He also asks whether recycling need always be so conscientious. Why, for example, do we flush toilets with water of dietetic quality?

The development of the artificial environment began slowly, depending on the chance discovery of suitable raw materials. To-day the range of available materials is very great and the pace of environmental change correspondingly rapid.

In a final chapter, Professor James K. Feibleman examines the emergence of civilization, showing how the invention of tools helped both to create and to control the artificial environment. The inheritance of acquired artifacts complements genetic inheritance, and does more; genetically-determined characteristics pass only from parent to children, but material culture can be transmitted to an entire generation.

As Professor Feibleman explains, there has been no progress in motivation. Man is equally effective at killing bodies and saving souls. The development of the environment has sharpened the conflict between aggression and affection. But there is no turning back. A return to nature, i.e. to a state of dependence on artifacts which do less to disturb the environment, is not possible for the millions, even though it can be achieved by the few.

JOHN LENIHAN
WILLIAM W. FLETCHER

Contents

CHAPTER ONE

THE MAN-MODIFIED CLIMATE OF TOWNS

TONY J. CHANDLER

THE URBAN ENVIRONMENT PROVIDES THE LIFE FRAMEWORK FOR A LARGE AND growing proportion of an expanding world population. Already about one-third of the world's peoples live in settlements of 5000 or more persons, and one-fifth in cities with a population in excess of 100 000. By the end of the century it is estimated that three out of every five people will live in towns.

The buildings in these urban centres are, with varying degrees of success, designed to modify the external climate in such a way as to bring internal conditions within an acceptable range of tolerance and comfort. But in doing this, the architect and town planner inevitably and often unconsciously change atmospheric properties outside, as well as inside, the buildings.

In substituting "artificial" urban units for natural shapes and surfaces, man has changed the physical and chemical properties of the air between and above the buildings. Each urban unit, each wall, each roof, each pavement, courtyard, street or park creates above it a climatological sheath with which it interacts. But these physical and chemical interactions serve to modify the atmosphere not only in the immediate vicinity of the buildings, but also throughout the entire built-up area and to some degree and in some respects, well beyond the city boundary. The urban dweller therefore spends most of his life in an "artificial" or substantially man-modified climate, indoors and out.

The changes effected upon the air by buildings and building groups are among the most radical induced by man upon the climate of the lower 600 m or so of the atmosphere known as the *boundary layer*. All the meteorological elements are changed, many by amounts sufficient to rank them as of major architectural, planning, social, medical and economic importance. The effects of the urban surface's complex geometry, the shape and orientation of individual buildings, the thermal and hydrological proportions of building, road and other urban fabrics, the heat from metabolism and various combustion processes operating in the city, and the pollution released from a variety of sources to change the chemistry of the city air, all combine to create a climate quite distinct from that of extra-urban areas.

It will be useful to consider each of the major meteorological elements in turn, analysing the way in which they are affected by buildings, building groups and entire settlements.

Airflow

Airflow in urban areas differs in nearly all respects from that above the aerodynamically smoother and generally cooler surfaces of rural areas. Because of these differences, a number of fundamental changes ensue. The shape of the mean horizontal wind profile in the boundary layer above urban areas is altered. Wind speeds and directions are changed in space and time, and the turbulence spectrum is quite different; and because the wind responds in such an intimate way to the detailed geometry and thermal distributions of urban areas, each city's pattern of airflow is, to a large degree, unique.

Because of the complexities and uniqueness, it is extremely difficult to find anything worthy of the term *representative site* for urban wind observations, and the extrapolation of measurements in one city to conditions in another has to be done with care. But although the variations from one place and one time to another may be considerable, this does not mean that generalizations have no value and that models of airflow have no role in building and urban design. Provided that the size of the variance is realized, generalizations or models of atmospheric behaviour in cities have an important part to play in proper urban design, just as they do in many other fields of environmental science.

J. K. Page has recognized three layers of urban airflow:

(*a*) A surface boundary layer close to the ground and between the buildings where the highly turbulent airflow is controlled largely by the aerodynamic shape of nearby buildings.
(*b*) A somewhat less-turbulent planetary boundary layer which extends from roof level to the level of the gradient wind, and which is more-or-less free from the influence of surface friction. The depth of this layer increases as air flows over the city.
(*c*) Surface-friction-free streamline flow above the boundary layer.

Wind profile

Wind speeds in the lower 40 m or so of the atmosphere, known as the *surface boundary layer*, are made extremely turbulent by the roughness of the earth. Owing to frictional drag with the surface, wind speeds also increase sharply with height, and the wind profile over a uniform extensive surface (conditions which, however, are unlikely to be properly met in an urban area) is normally assumed to approximate to a logarithmic profile:

$$\bar{u}_z = \frac{u_*}{k}\log\frac{z}{z_0}$$

where $\bar{u}_z$ is the mean horizontal wind speed at height z
u_* is the friction velocity
k is von Karman's constant and
z_0 is the roughness length.

But in addition to the other limitations upon the use of this formula in urban areas, it has been pointed out that it assumes a balance between the production and dissipation of turbulent kinetic energy (the energy of motion) which is again unlikely in a built-up area.

Above about 40 m, the logarithmic law no longer seems to represent conditions fairly, even in rural areas, and the increase of speeds from this level to the top of the boundary layer, away from the earth's frictional drag, is more accurately given by a power relationship:

$$\bar{u}_z = \bar{u}_1\left(\frac{z}{z_1}\right)^a$$

where $\bar{u}_1$ is the mean horizontal wind speed at height z_1 and
a is an empirical power exponent.

The power-law relationship seems to be as valid and more easily used than many other statistical expressions of the velocity profile, and it is the one most commonly used in calculating winds at a series of heights in the atmospheric boundary layer. It has been suggested, however, that in cities, heights should be calculated, not from the ground but from a level approximating to general roof height in the surrounding area—a principle known as *zero-plane displacement*. Below this height, i.e. in the streets, squares and courtyards of urban areas, winds are highly turbulent and are often organized in eddies, so that the wind profile has little meaning.

Investigations in a number of cities have yielded a range of values for the exponent a in the power-law relationship. These are listed in Table 1.1.

Table 1.1 Estimates of power-law exponents in different cities

City	*Upper limit of investigation* (m)	*Exponent*	*Reference*
Paris	305	0·45	Eiffel (1900)
Leningrad	149	0·41	Ariel and Kliuchnikova (1960)
New York City	381	0·39	Rathbun (1940)
Copenhagen	73	0·38	Jensen (1958)
London (U.K.)	183	0·36	Stellard (1967)
London (Ontario)	40	0·36	Davenport (1967)
Kiev	179	0·35	Ariel and Kliuchnikova (1960)
Tokyo	250	0·33	Soma (1964)

Thus of all the formulae proposed for modelling the increase in mean horizontal wind speed with height above an urban area, the power law seems to give as reasonable and certainly as easily obtainable results as others, though several researchers have noted the difficulties in deriving reasonable values for the zero-plane displacement.

Studies have shown that the power-law relationship gives, height-for-height in the boundary layer measured above the ground, lower wind speeds over a city than rural surfaces, and lower speeds here than above a low-friction sea surface. The rate of increase of wind velocity with height (wind shear) is, however, greatest above the city. With an exponent of 0·40, winds at 170 m above ground, for instance, are equal to 90 per cent of the Gradient Wind (blowing at the top of the boundary layer) over the sea, 75 per cent of the Gradient Wind over rural areas, but only 60 per cent of the Gradient Wind over a city (figure 1.1).

The velocity profile must, of course, take some time to adjust to changes in surface roughness conditions. It has been estimated that for winds blowing from rural to urban areas, the profile does not become adjusted to the increased surface rougness for just over 1 km at a height of 30 m and more than 8 km at a height of 100 m downwind of the rural/urban boundary. With winds blowing from urban to rural areas, the distances are about 3 km and 14 km respectively, so that it appears to take twice the distance for adjustment in winds blowing out of an urban area as for those blowing in.

There are several important implications. First, the *surface of adjustment*, defined as the top of the adjusted layer, slopes upward from the rural/urban boundary with a slope which is initially about 1:32 and then shallows to about 1:80, which roughly equals the slope of the deepening boundary layer as a whole. Secondly, the adjustment distance at 100 m for winds blowing from country to city is greater than the width of many towns, which means that in these cases winds will never achieve the true

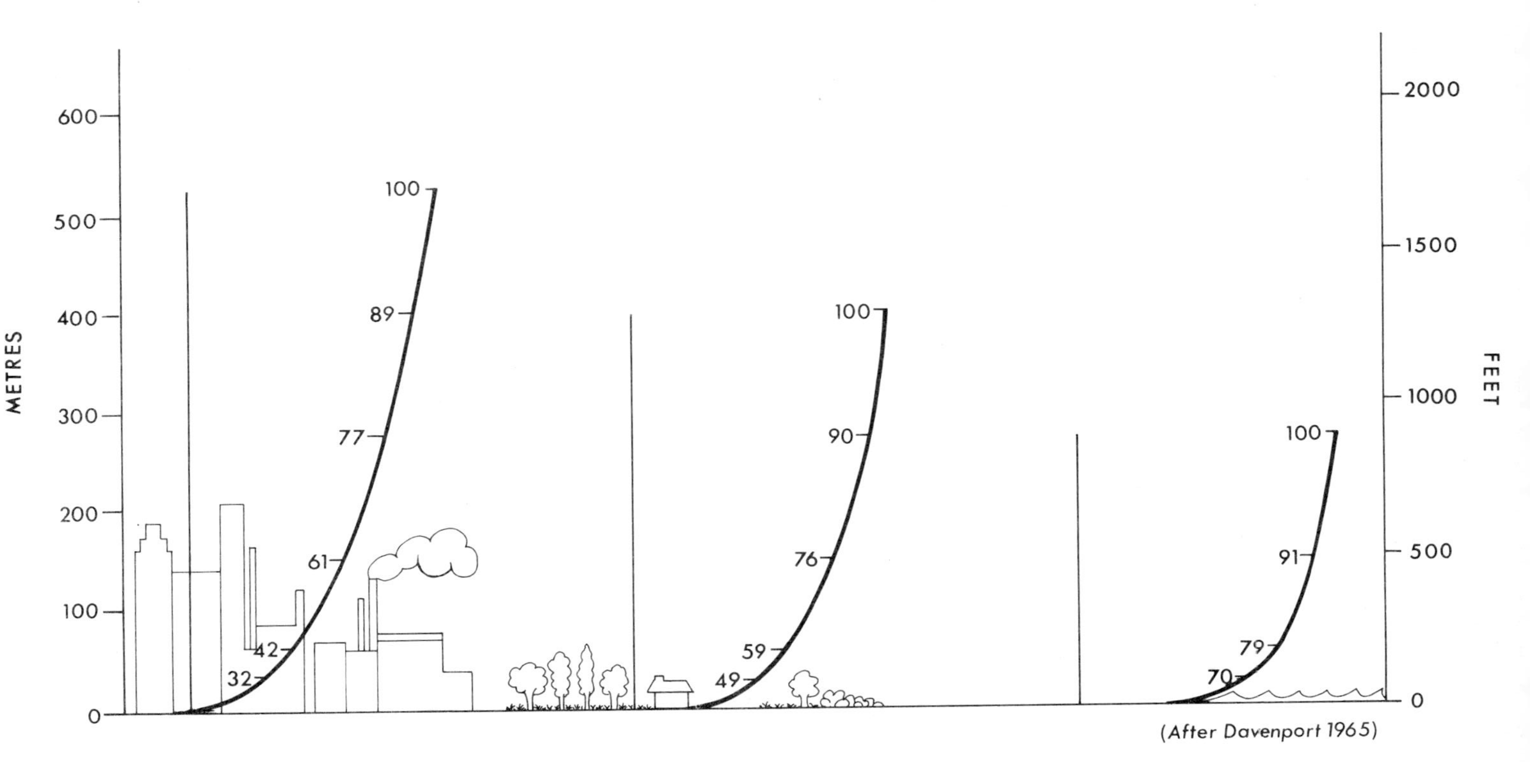

Figure 1.1 Typical wind velocity profiles above urban, rural and sea areas. The numbers give, for each level, the mean horizontal wind speed as a percentage of the Gradient (surface-friction-free) Wind.

urban wind speed for this height. These features have obvious implications for the design of tall buildings in urban areas.

The depth of the boundary layer also responds to the degree of surface roughness and is much deeper over urban than over rural areas. Representative depths are as follows:

	Representative depth of the boundary layer in metres
Flat, open country	275
Woodland	400
Towns and city suburbs	400
City centres	425

What is interesting and, indeed, highly relevant to building engineering, is the question of how rapidly the boundary layer adjusts itself to changes in surface roughness. How quickly does it deepen when crossing the rural/urban boundary, and how rapidly does it shallow again when leaving the downwind side of the city? There is, unfortunately, only limited theoretical and observational evidence by which to answer these questions.

It has been suggested that on nights with a well-developed temperature inversion (increasing temperature with height) at the top of the atmospheric boundary layer, the airflow aloft might become "decoupled" from that below. Rising over the generally domed upper surface of the mass of warm air over the city, known as its *heat island*, the winds might accelerate as they do over hills in their path. It is therefore interesting to note that unusually strong winds have been found at heights of about 200 m above several cities.

Because of differences in the depth and intensity of turbulence and in the consequential vertical transport of momentum, winds near the ground tend to be strongest by day and weakest by night, whilst at higher levels the reverse is generally true. Such differences are likely to be important over the height range of many modern buildings.

Wind speed

Because of increased surface friction, average wind speeds are reduced in cities in comparison with those in country areas. In autumn, winter and spring, for instance, when regional winds tend to be strong, speeds in central London are reduced by eight, six and eight per cent respectively, whilst in summer, with lighter winds, there is little or no difference in mean urban and rural speeds. The overall annual reduction in wind speed in central London, for all winds, is six per cent, but for winds of more than 1·5 $m\ s^{-1}$ the reduction is 13 per cent.

But although average wind speeds are less in urban than in nearly rural areas, on individual occasions the relationship may be reversed. Independent analyses of London and New York data have both shown that the rural/urban wind speed difference was a function of the regional near-surface wind speed and wind profile. More significantly, when winds were light, near-surface speeds were greater in the built-up areas than outside, whereas the reverse relationship existed when winds were strong. The critical threshold speed at which the sign of the urban/rural speed difference was reversed in the London case was 3 to 5 m s^{-1} as recorded at London Airport on the western fringe of the city. The relationship between speeds at London Airport and at Kingsway in central London has been found to depend on the season, time of day and regional wind speed (Table 1.2).

Table 1.2 Average wind speeds at London Airport (on the western fringe of the city) and its excess above that at Kingsway (in central London) in m s^{-1}

	0100 GMT		1300 GMT	
	Mean speed	*Excess speed*	*Mean speed*	*Excess speed*
December–February	2·5	−0·4	3·1	0·4
March–May	2·2	−0·1	3·1	1·2
June–August	2·0	−0·6	2·7	0·7
September–November	2·1	−0·2	2·8	0·6
Year	2·2	−0·3	2·9	0·7

The acceleration of speeds in central London has been shown to be mainly a night-time phenomenon, brought about by the strong downward transport of momentum induced by mechanical turbulence above the city, which at times of light winds and a sharp wind speed profile (fairly common night-time occurences) more than compensates for the increased surface friction of the buildings.

R. D. Bornstein *et al.* showed that in winds above New York of less than 4m s^{-1}, there was a 20 per cent increase in speed over the city, the greatest increases occurring in winds of less than 1·3 m s^{-1}.

Wind accelerations have also been monitored by J. K. Angell *et al.*, using tetroon* flights above Columbus, Ohio. They found the size of the acceleration to be dependent upon the strength of the nocturnal inversion and registered a deceleration of the wind downwind of the city centre, as had Bornstein in New York. Lee, on the other hand, was unable to find convincing proof of this suburban deceleration in his studies in London.

*A tetroon is a small constant-volume balloon.

Turbulence

The spatial and temporal variations in wind speed and direction are much stronger in city than in country areas, but very few detailed studies have been made of turbulence in urban areas, and the results in any case would be particular to the detailed surface geometry of individual cities.

One of the few studies available is that by I. R. Graham in Fort Wayne, Indiana. He showed that above his city, set in an almost perfectly flat featureless plain, the intensity of turbulence as measured by the standard deviations of the azimuth angle of the wind, was closely related to the density and form of the development. The pattern of turbulence was concentric to the city centre, where standard deviations were more than twice as large as in the surrounding country.

Observations taken on tall meteorological observation towers show gustiness to decrease fairly rapidly with height. But the increase of "instantaneous" gust speeds with height is slower than the increase in mean velocities, so that whilst gust speeds in the surface boundary layer are generally in excess of mean wind speeds, at higher levels they are more similar.

Wind direction

In the lowest part of the boundary layer, between and immediately above the buildings of a city, airflow is very turbulent and wind direction is variable in both space and time. The surface orography upon which the city is built also exercises a unique control in each case. In spite of these complexities, there are patterns of change in the three-dimensional airflow in urban areas, and these are sufficiently consistent from one city to another to constitute a real urban model.

Above the turbulent eddies which characterize airflow around buildings, winds blow at an appreciable angle to the isobars because of the powerful surface friction. Measurements of this angle in a number of cities vary from 15° at times of strong instability to more than 40° during temperature inversions. Several investigators have shown the tendency for anticyclonic turning of the streamlines of airflow in the lowest levels of the air over cities by day, followed by a cyclonic recovery downwind, and explained this as a consequence of frictional turning, vertical mixing and a mesoscale high-pressure distribution at the top of the heat dome. Johnson and Bornstein found evidence of cyclonic curvature over New York in strong winds and anticyclonic curvature in light winds.

At times of strong stability and a marked heat island, however, the heat dome over a city may act as an obstacle diverting airflow around either side and over the top, with the consequential accelerations.

Many tetroon flights over American cities have traced strong rising and sinking air currents over built-up areas by day. In studies by A. W. Hass over New York City, for instance, a mid-afternoon flight showed a rise of over 800m above the warm physical obstacle of Manhattan, with almost vertical sinking again over the East River. Studies by J. K. Angell *et al.* showed that the strength of the upcurrents and down-draughts over Columbus, Ohio, were very dependent upon the stability of the urban air.

Local airflows

At times of calm or very weak regional winds, urban heat islands will set up a country-city airflow system in many ways similar to the sea-breeze. Studies in London, Leicester, Frankfurt, Bonn, Asahikawa (Japan), Louisville, and New York, have all provided field evidence for the surface link in this type of airflow. This consists of a cool inflow of a few metres per second towards the centre(s) of warm air. These shallow centripetal flows are rapidly decelerated by surface friction in the suburban fringe and, for this reason, they rarely penetrate more than a few kilometres into the city. Here they rise, probably in a rather diffuse zone of uplift, to join an outblowing centrifugal flow from city to country again at a higher level. The cellular system is then completed by sinking over the country areas around the city. The surface inflows are pulsating in character, for, in blowing across the urban margins, they erode the shape thermal gradients of the heat island which were their cause. They therefore die down until a new cliff-like edge to the heat island is built up, which once more generates thermal country-wind across it.

These country breezes, blowing across the margins of the heat island, help to sharpen the edge of the urban pollution dome at times of calm or light winds.

Air pollution

Man has, until recently, largely ignored the fact that locally and perhaps globally, he has overloaded the natural self-cleansing capacity of the atmosphere so as to produce very high and perhaps dangerously high concentrations of air pollutants, many emanating from combustion processes of one type or another. Because man's socio-economic organizations have concentrated his waste-forming activities in towns, it is here that the most serious problems are generally, though not exclusively, to be found.

It must be remembered that all or nearly all pollutants (the exceptions are mainly rather specialized radioactive substances) exist naturally in the atmosphere. Man's role has been to increase concentrations locally or through time.

Emissions, transport and abstraction processes are complex and the subject of intensive and urgent studies in most countries, by a variety of scientific disciplines.

Nature of pollutants

The pollutants of greatest concern in urban areas are those released as waste products from combustion processes of one sort or another. Among those which have attracted the greatest attention are the following: particulates including various metals, oxides of sulphur, carbon monoxide, carbon dioxide, oxides of nitrogen and hydrocarbons. These are emitted from a variety of sources in urban areas, including tall stacks, industrial "process" emissions, and low-level area and mobile sources. Released in differing amounts and at various speeds and temperatures from a multiplicity of fixed and moving sources, it is extremely difficult to compile detailed inventories of the patterns of emission in urban areas. Add to this the complexity of the meteorological controls upon diffusion, chemical changes and abstraction rates, and we begin to understand the limitations which presently exist upon knowledge, though the enormity of the national, regional and international scientific effort within the field of air pollution in recent years has undoubtedly greatly improved our understanding of the problem.

In 1969, estimates were made of the contribution of various sources to the total weight of air pollutant emissions in the United States. Fuel consumption from stationary sources contributed 16 per cent of the total; transportation 51 per cent; industrial processes 14 per cent; garbage incineration 4 per cent, and other sources 14 per cent. Different countries have developed different attitudes to each of the different pollutants and set varying priorities for their emission control. In most European countries, smoke and sulphur oxides have been accorded first priority; in the United States, sulphur oxides were rated only eighth in order of priority, oxides of nitrogen and hydrocarbons coming first and second respectively. Low-latitude cities, with very different combustion patterns, would presumably generate different air-quality priorities again.

Air pollution is not, of course, a uniquely urban problem, although the greatest mass of pollutants are generated in towns. These pollutants are then carried distances varying from a few metres to thousands of kilometres before being returned the earth's surface, depending upon the height and other physical properties of emission, the character of airflow and temperatures in the atmospheric boundary layer, precipitation and the aerodynamic and chemical properties of the pollutants. Though nowhere entirely escapes the problem, cities are clearly the most polluted areas. The larger the city, the higher in general are the concentrations of the main pollutants, although the relationship between concentration and city size is by no means simple or constant. Much depends upon the precise location of the monitor(s) in relation to sources of emission. The gradients of concentration away, for instance, from individual stacks or main highways, are often steep, and for this reason the records of individual or small numbers of monitors must be treated with caution.

Concentrations of pollutants in space and time

Before the advent of individual and collective environmental consciousness, leading to the gradual improvement in many aspects of air quality, man had wantonly discharged vast quantities of waste products into the air from house and factory chimneys and plant, and from quarries, cars, lorries, trains and ships. These still plague many cities but, because local concentrations are derived in large part from nearby sources, the pattern of pollution and, more particularly, of smoke and vehicle emissions in cities, are closely related to the individual pattern of emissions. This is particularly true at times of stability and light winds when concentrations in urban (unlike rural) areas are highest. For this reason, it is less easy to make meaningful generalizations about the distribution of pollutants in cities than about most other meteorological elements; concentrations vary over very small distances and very short periods of time and so produce complicated spatial and temporal patterns of pollution, as individual to each city as the distribution of its emissions. W. Bach, for instance, studied the turbidity coefficient (an integrated measure of the vertical pollution load) in various land use districts of Cincinnati and showed the frequency of polluted air (having a turbidity coefficient of 0·21 or greater) to increase from a city park to the suburbs and then in ascending order of percentage frequency, through city residental areas and the city centre to an industrial area. The industrial site experienced almost exclusively "polluted air" whilst the park had predominantly "clean air" (turbidity coefficient of 0·10 or less) over the same summer period.

Land also affects the form of the diurnal variation of aerosol concentrations, residential areas showing a typical morning and evening peak at times of heavy emissions and moderate or strong stability.

Patterns of sulphur dioxide, the major gaseous pollutant of most mid-

and high-latitude cities, are much less complicated than those of smoke. This is because the height of emissions is generally greater and the aerodynamic properties of a gas are different from those of an aerosol. In consequence, sulphur dioxide is capable of drifting further than smoke and does not show the latter's sharp gradients within cities or precipitous fall-off in concentrations near the margin of the urban area. For the same reasons, sulphur dioxide can drift further away from its source than can smoke. Much industrial emission is from tall stacks which help to diffuse the pollutant efficiently so as to produce low ground-level concentrations, especially in their immediate vicinity. This is because ground-level concentrations of pollution are, on average, inversely proportional to the square of the effective height of the emission, with the greatest concentrations some distance from the stack. Domestic smoke concentrations, on the other hand, are released (mainly in winter) at lower temperature and from much lower levels than industrial pollutants, and are caught in the eddies around buildings and across streets so as to most severely pollute those areas in the immediate vicinity of the sources. This close relationship between emissions and concentrations is fundamental to the success of local smoke control orders in residental areas, allowing considerable local improvements to result from actions to reduce emissions.

The closeness of the relationship will, of course, vary from one city to another and through time, and will depend upon prevailing meteorological conditions; high ground-level concentrations of smoke and sulphur dioxide are associated in most towns with light winds and a stable atmosphere within the mixing layer up to the effective height of emission.

The lateral drift of smoke across cities by the prevailing wind is important, but much less so than has often been supposed. Except at times of light winds and stable atmospheres, smoke tends to spread vertically more than laterally, the reason being the wide-angled diffusion of smoke from low-level sources, including downwash in the wake of buildings, and the often turbulent nature of prevailing winds. The lateral drift of sulphur dioxide across a city is much greater than that of smoke, with substantial rises in ground-level concentrations in winds that have crossed the city.

Although the greatest contribution to ground-level concentrations will generally come from nearby emissions, additions can sometimes be made by sources far beyond the city boundary. Smoke and sulphur dioxide can drift hundreds, sometimes thousands, of miles from their source. They often do so several hundred metres above the earth frequently becoming trapped and concentrated under a temperature inversion, before being brought down to the ground, again by the increased turbulence of a city lying in the path of a plume. For example, concentrations of smoke and sulphur dioxide in eastern Reading, 64 km west of the centre of London, are often significantly enhanced by downwash from a high-level drift from London at times of light easterly winds.

Concentrations of pollution such as carbon monoxide, oxides of nitrogen, and hydrocarbons from mobile sources fall off very rapidly in streets leading away from the main traffic highways. In streets orientated at right angles to wind, concentrations are affected by eddies which form in these situations with a downdraught of relatively clean air on the lee side and the highest concentration of vehicular pollutants on the lower windward side of the street chasm. The extra turbulence produced by cars on major urban highways may help to keep concentrations of pollution down at times of peak traffic flows, and there is some evidence of moderate increases in pollution concentrations occurring soon after (rather than during) the morning and evening peak traffic periods.

Radiation and sunshine

Because of the blanket of pollution which, to a greater or lesser extent, shrouds all urban areas, radiation receipts at the ground are reduced, often significantly and especially at times of low solar elevation. It is likely that most of the attenuation is by absorption, since most scattered radiation will be directed forwards. It has been estimated that over heavily polluted areas, absorption can be sufficient to cause temperature rises of 10°C per day. This urban effect is particularly noticeable in high-latitude cities during the early morning and late evening in winter.

An attenuation of the solar radiation by between 10 and 30 per cent has been reported from many cities. H. J. de Boer, using two years of global solar-radiation measurements at six stations in and around Rotterdam, showed that the centre of the city received 3–6 per cent less radiation than the suburbs, and 13–17 per cent less than the country. From November to March, and before the reductions in air turbidity following the U.K. Clean Air Acts, many smoky British cities received between 25 and 55 per cent less radiation than nearby rural areas. In central London the loss of sunshine amounted to about 270 hours per year, there being a reduction of more then 50 per cent in December. With the more recent improvements in air quality, I. Jenkins has shown that winter sunshine receipts in London have increased by 50 per cent, being as much as 70 per cent in January. On a shorter time scale, investigations in many cities have demonstrated a weekly cycle of radiation with higher amounts at weekends when industrial emissions of pollution are generally less. C. L. Mateer has shown that in Toronto, Canada, solar-radiation receipts were 2·8 per cent greater on

Sundays than the average for other days. Furthermore, from October to April the Sunday excess was 6·0 per cent, but during the remaining months it was only 0·8 per cent.

It is also important to note that the attenuation of radiation by pollution varies significantly with wavelength. C. H. Maurain has reported a 100 per cent decrease in the ultra-violet wavelengths between the centre and outskirts of Paris. In Leicester (England), A. R. Meetham demonstrated that ultra-violet radiation in the range 0·314–0·355 μm was reduced by 30 per cent in winter but by only 6 per cent in summer. Clear-day attenuation of ultra-violet radiation between Mt. Wilson (5350 ft) and central Los Angeles has been measured by J. S. Nader at 14 per cent on clear days, rising to 58 per cent on smoggy days; and R. Stair has shown that this increases to 90 per cent or more under extreme conditions. In the infra-red wavebands, W. T. Roach found that up to 35 per cent of solar radiation over southern England was absorbed by pollution in the lower 1000 ft of atmosphere.

There have been very few comparative studies of urban and rural long-wave radiation amounts, but 6–8 per cent higher daytime values have been recorded over cities such as Cincinnati.

As with other elements, only a small number of studies have been made of the variation of radiation with height over urban areas, though most of these have shown a layered structure of temperature and pollution over the city which significantly reduced the amount of radiation reaching the surface.

Temperature

Within the field of urban climatology, more attention has been paid to the study of temperatures within cities than to any other meteorological element with the possible exception of pollution.

The warm air which frequently covers built-up areas, particularly by night, is known as a *heat island*. Heat islands have been charted in a large number of cities, most of them in mid-latitudes.

Horizontal temperature field

Urban-rural temperature differences stem from the integrated contrasts between town and country in each of the terms of the energy balance. Many of these elements undergo characteristic diurnal, seasonal and synoptic changes in their values so that the overall urban-rural temperature difference varies not only from one city to another and through time, but also from one part of a city to another. This is because local air temperatures respond to the make-up of the internal land-use elements, so that areas of similar urban development give rise to fairly uniform air temperatures and are separated from other discrete development areas, often by very sharp thermal gradients.

In general, urban heat islands are most intense by night when the urban-rural temperature difference has frequently been measured as 5°C and may reach 11°C. Maximum heat-island intensities are usually attained a few hours after sunset, mainly as a result of strong rural cooling. Later in the night, urban cooling rates often slightly exceed those in rural areas, and the urban-rural temperature difference grows less. After dawn, the vegetation-covered soils of rural areas, with a relatively low thermal capacity and fully exposed to solar radiation, warm faster than air in city streets, so that the urban-rural temperature difference further decreases or even reverses to form an urban "cold island".

Table 1.3 gives the mean temperatures in and around London for the period 1931–60.

Table 1.3 Average annual temperatures in and around London, 1931–60

	Average height (m)	*Maximum* (°C)	*Minimum* (°C)	*Mean* (°C)
Surrounding country	87·5	13·7	5·5	9·6
Suburbs	61·9	14·2	6·4	10·3
Central districts	26·5	14·6	7·4	11·0
Heat-island intensity				
Central districts		0·9	1·9	1·4
Suburbs		0·5	0·9	0·7

These average values are probably fairly typical of most medium to large towns in mid-latitudes. Much more work needs to be done before we can generalize about settlements in other climatic regimes, although such studies are now becoming more common.

By day, heat islands are much less intense; indeed, cold islands with lower city than rural temperatures are quite common. In central London, for instance, daytime cold islands occur on about one day in three on average, and one day in two from February to April; by night the frequency is less than one night in five with a clear winter peak. Part of the explanation of lower daytime temperatures within cities than outside is the thermal-lag effect of areas such as city centres having a high thermal capacity, but also important in this respect is the shading of city streets, gardens and courtyards by tall buildings. For this reason, some investigators have shown the highest daytime temperatures to occur in a ring around Central Business District (downtown) areas of cities.

Detailed studies in many cities have demonstrated the complex vaguely-concentric pattern of isotherms in cities, though with local peaks and hollows in the temperature surfaces reflecting changes in urban building forms and densities (figure 1.2). In spite of individual differences, certain features are common to most heat islands. One is the close correspondence

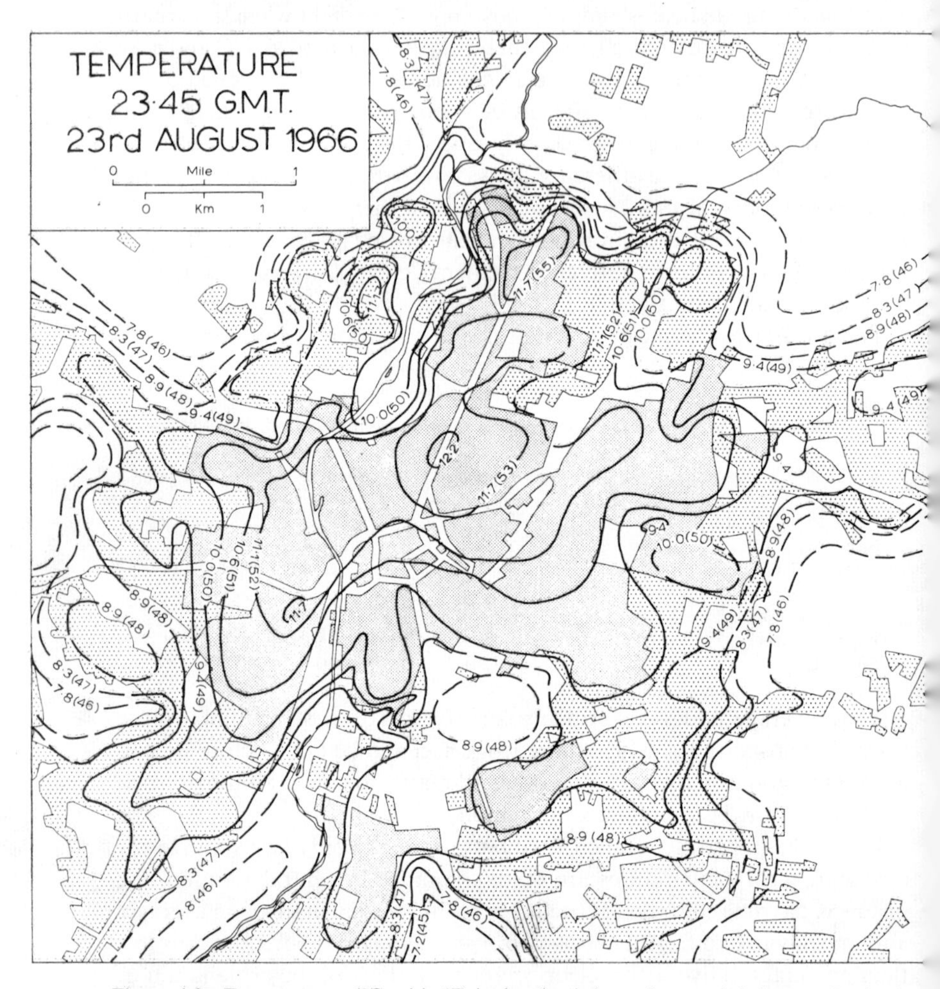

Figure 1.2 Temperatures (°C with °F in brackets) in and around Leicester, England (population 287 000) at 23.45 GMT on 23 August 1966. The built-up area is classified, rather subjectively, into areas of contrasted building densities. One, generally found in more central areas, is composed of mainly late eighteenth, nineteenth and early twentieth-century buildings, both domestic and industrial, with few open spaces. Here, building densities are high and roads narrow. The other, typical of the outer suburbs, is mainly post-1918 development of much lower building density containing many parks, allotments, gardens and other open spaces.

found, more especially on calm clear nights, between the pattern of temperatures and the urban morphology. Highest night-time temperatures are generally found in the densely built-up areas of city centres. It is here that the surface areas and the thermal capacities of the buildings are greatest, and diffusion is weakest in the street and courtyard canyons and cavities between buildings. The absolute values and pattern of temperatures in a city are profoundly influenced by site conditions and by the prevailing weather conditions. In general, although the usual pattern of isotherms is roughly concentric, reflecting the annular arrangement of most urban forms (figure 1.2), on individual occasions other than during absolute calms, the peak temperatures will be displaced somewhat downwind of the areas of densest building development by the advection of cold air into the windward side of cities.

As a result of changes in meteorological parameters, especially stability, wind speeds and cloud amount, heat islands will constantly change their form and maximum intensity, even within an individual settlement. As a result, mean urban-rural temperature anomalies, which often average between 1·5° and 2°C for night-time conditions over a broad spectrum of settlement form and size, cover a very wide range of individual values.

Differences between cities are probably due not only to contrasts in their urban morphology and the relative importance of the liberated heat of combustion, but also to differences in the general climates of their regions, more particularly in relation to seasonal changes in wind speed, cloud amounts and the incidence of inversions. These have also been held responsible for allegedly more intense heat islands in continental than in maritime climates.

Although night-time peaks of heat-island intensity are almost universal, there are large and largely unexplained regional variations in the seasonal pattern of values. Many cities such as those in the United Kingdom and western North America record a clear summer maximum in mean intensities; others, such as many in Japan, show a winter peak; while others, such as several in central Europe, Scandinavia and Central America, record no clear seasonal rhythm in values.

Because of the dependence of heat-island intensity upon a series of variable parameters, including cloud amount and wind speed, year-to-year changes are also apparent in mean intensities.

But what has emerged in recent studies of temperatures and other elements in urban areas, is the close correspondence between local values and nearby urban development forms and densities. For instance, on individual nights of stability, clear skies and calm, areas of similar urban development in Leicester (population 270 000) and London (population 8·25 million) have been shown to record the same excesses of temperature above those of nearby rural areas. City size thus seems to be relatively unimportant, the intensity of the immediate urban development being the more important determinant of the strength of the heat island. At other times and over a long-time average, there does seem to be a close relationship between city size and heat-island intensity. J. Murray Mitchell has shown that during the first half of this century, city temperatures increased faster than those of rural areas, though it is possible that this was because of trends in global climate rather than city growth. H. Dronia made similar observations using 67 paired urban/rural stations around the world, but T. J. Chandler showed that there was no such consistent trend in London's temperatures which responded to year-to-year fluctuations in the primary meteorological controls upon heat-island intensities. T. R. Oke, on the other hand, has demonstrated a close logarithmic relationship between the average urban-temperature excess and city size, measured as population, for North American and European cities.

The size of the city may be relevant to several parameters of the urban heat balance, but two of its main effects are likely to be first through the roughness and fetch (distance of urban traverse) influences upon airflow (and therefore upon wind speed and turbulence at the city centre) and, secondly, through the relationship that commonly exists between city size and the intensity of development in the central business district. Large cities, unlike smaller towns, are likely to have massive central building developments, and this alone will affect the peak intensity of their heat islands.

City size also seems relevant, through its control upon critical wind speeds, for the elimination of the heat island. T. R. Oke and F. G. Hannell have found a statistical relationship (correlation coefficient $+0{\cdot}97$) between this critical wind speed U and the logarithm of the population P (used as an index of the spatial extent of the city) in the form of the regressive equation:

$$U = -11{\cdot}6 + 3{\cdot}4P$$

For London (population 8·25 million) the critical speed for the elimination of the heat island was 12 m s $^{-1}$; for Montreal (population 2 million), 11 m s $^{-1}$; for Reading, England (population 120 000), 4·7 m s $^{-1}$: and for Palo Alto, California (population 33 000), 3·5 m s $^{-1}$. The authors then used this equation to estimate the smallest settlement that would form a heat island. When $U=0$, $P=2500$, but other evidence shows that small building groups and even individual buildings can produce measurable effects upon local temperatures. In much the same way, T. J. Chandler has noted a strong correlation between local heat-island intensities and the amount (roughly volume) of urban development within 500 m of the points of measurement.

Several authors have noted the surprisingly large heat islands that can sometimes develop over small settlements, provided their building densities are high. Others have studied the effects of parks and squares, noting the lower temperatures that are usually found in open spaces within cities, a feature of some importance in urban planning.

Vertical structure of temperature

Much less is known about the vertical than about the horizontal pattern of temperatures in heat islands. Meteorological towers, balloons and tall buildings carrying temperature recorders have been used in several cities, but perhaps the most detailed results have come recently with the use of instrumented helicopters.

All of the studies show that, because of increased thermal and mechanical convection, night-time inversions of temperature are less frequent, weaker, and higher over the city than over the country, although there is often little difference between the two environments by day. By night, the well-mixed mixing layer shrinks in depth but, except in the lower 50 m or so, still exhibits a weak stability structure. In his study of New York City, R. D. Bornstein recorded multiple elevated inversions over the surrounding country. Between the ground and heights of up to 300 m over the city, temperatures were generally higher than over the country and the lapse rate was pseudo-adiabatic. Above about 400 m, temperatures over the city were generally lower than those above the surrounding country. This "crossover" of urban and rural temperature profiles with colder air above the city than the country at a level of 300–400 m above ground has not been fully explained, but it is suggested that it may be related to long-wave radiation from a high-level pollution haze over the city. Other suggested causes are the adiabatic effects of vertical mixing and urban-induced cellular circulations linking city and country.

At times of clear skies and light winds, several authors have found evidence of an "urban heat plume" rising above the city and then spreading several miles downwind at intermediate heights. Upwind of the urban area, there is normally a strong deep radiation inversion, whilst over the built-up area, temperature lapse conditions prevail up to about 60 m above the ground. Downwind of the urban area a strong but shallow inversion commonly exists at the surface, above which there is often a weak lapse condition in the urban heat plume, followed at higher levels again by the upper part of the regional nocturnal radiation inversion.

With calm air, the envisaged model of the boundary layer is of an urban "dome" rather than "heat plume" rising downwind from the city. Within the dome, both heat and pollutants are well mixed, with steep thermal and concentration gradients near the upper limit of the dome.

Humidities

There have been relatively few comparative studies of the moisture content of urban and rural atmospheres, and the observational evidence available is not completely conclusive.

Much of the urban surface is sealed by concrete, asphalt and other impervious and semi-impervious materials, although the fairly large amount of open ground that still remains in most cities has sometimes been overlooked. Certainly a lot of rainwater is led quickly underground, but the amount of moisture absorbed by bricks and tiles has frequently been underestimated. Nevertheless, in spite of all these qualifications, there is no doubt that evaporation rates in cities are substantially less than from vegetation-covered soils. On the other hand, amounts of dewfall are almost certainly greater in rural areas, and the added moisture from combustion is greater in the city.

Contrasts in evaporation rates are likely to manifest themselves in differences in absolute humidity only when the atmosphere is calm. For this reason, average or monthly vapour pressure contrasts between urban and rural areas are likely to be very small. It has, for instance, been found that the annual vapour pressure in London is only 0·2 mb lower, on average, than at a nearby rural climatological station. On calm nights, however, the moisture content of the air above London and above another English town, Leicester (population 270 000) is highest in the centre of the city, being 1·5 to 2·0 mb higher there than above the surrounding country (figure 1.3). This is probably explained in terms of the low rate of diffusion of air "trapped" between tall buildings so that it retains its characteristically high daytime humidities.

Relative humidities, being a function of prevailing temperatures, are normally found to be inversely related in towns to the local intensity of the urban heat island, and average annual urban-rural differences of about 5 per cent have been widely reported. On individual calm clear nights, the patterns of relative humidities in cities are closely related to the detailed form of their heat islands, which in turn depend upon the distribution of urban building densities. Relative humidity differences between town and country can, on these occasions, amount to between 20 and 30 per cent.

Few studies have been made of the vertical gradients of humidities over urban areas. One of these by R. D. Bornstein, A. Lorenz and D. S. Johnson, reported a night-time excess of absolute humidity of about 4 per cent to heights of 500–700 m above the built-up area of New York City, the greatest differences occurring during the early morning when the vertical mixing was strong.

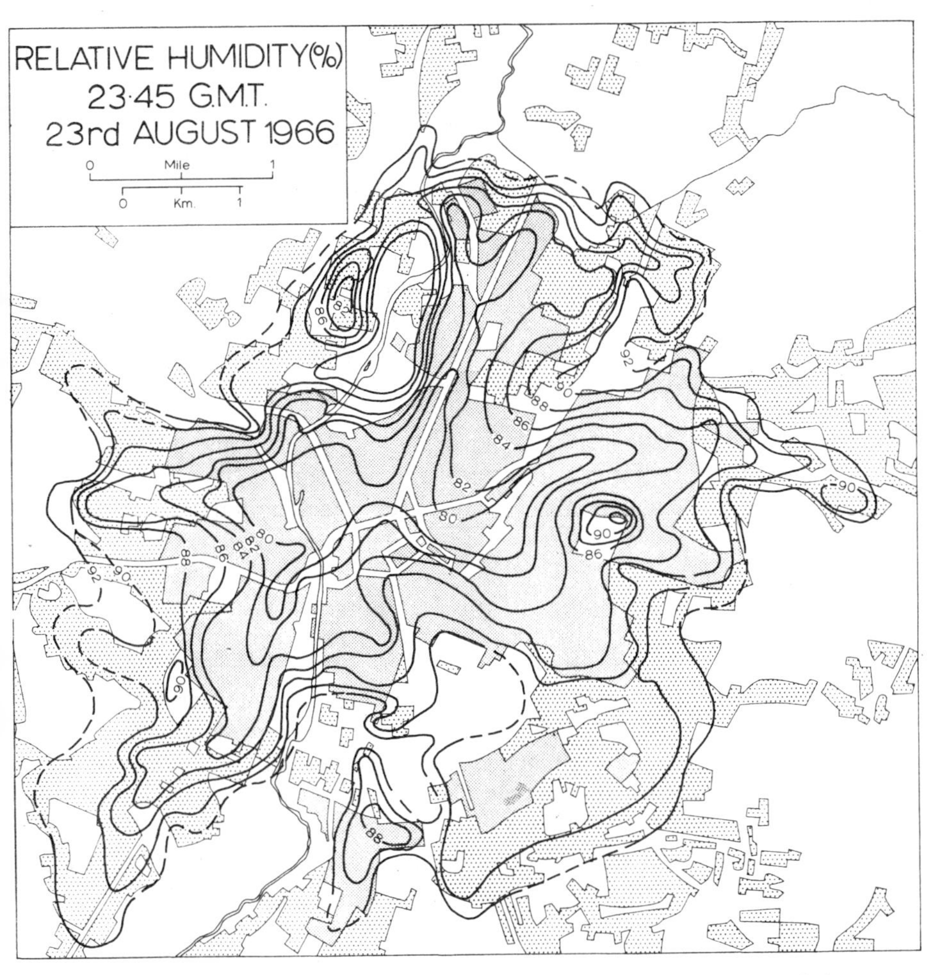

Figure 1.3 Relative humidities (per cent) in and around Leicester, England (population 287 000) at 23.45 GMT on 23 August 1966. The built-up area is classified, rather subjectively, into areas of contrasted building densities. One, generally found in more central areas is composed of mainly late eighteenth, nineteenth and early-twentieth century buildings, both domestic and industrial, with few open spaces. Here, building densities are high and roads narrow. The other, typical of the outer suburbs, is mainly post-1918 development of much lower building density containing many parks, allotments, gardens and other open spaces.

Precipitation

In contrast to the other near-surface meteorological phenomena, much of the literature on urban precipitation is primarily concerned with establishing whether or not the urban area has any effect at all on the precipitation distribution. At first sight it seems plausible to postulate that because of increased pollution (with active condensation and freezing nuclei, as well and more active thermal and mechanical turbulence, and the water vapour from combustion processes) there will not only be increased cloud but also more precipitation above cities. Several authors have produced theories and field evidence for increases of precipitation in a number of cities in Europe and the United States.

Simple comparisons of precipitation inside and outside cities prove very little, because of the well-known temporal and spatial fickleness of the element, the relatively poor sampling inherent in normal rain-gauge measurements, the drift of city pollutants into rural areas, and the important orographic controls upon precipitation exercised by the complex underlying terrain of most cities. Also, many recent studies have emphasized the time element in precipitation processes and have drawn attention to the common displacement of positive anomalies downwind of city centres. The major enhancement of summer rainfall in the St. Louis region, for instance, often occurs downwind of the city where it is most likely affected by the urban heat plume. Whether there is enhancement or not also clearly depends on the synoptic situation, the greatest enhancement over St. Louis occurring in air masses free from synoptic disturbances.

Apart from the highly individual meteorological circumstances behind the famous positive precipitation anomaly in La Porte, a small settlement east of Chicago-Gary, most increases of precipitation over urban areas in the United States average between 5 and 8 per cent in annual totals (being greater in the cold than warm season) and between 17 and 21 per cent in the number of summer thunderstorms. Similar increases have been reported for Munich, Nuremburg, Budapest, London, and Bombay.

Cities are unlikely to make very substantial differences in the frequency of snowfall, for snow falls at times of the year and in meteorological conditions generally inhibitive to strong heat islands which would otherwise act both as a trigger to instability and a means of melting the flakes. Nevertheless, a few instances have been reported where snowfall over an urban area was lighter than at nearby rural locations. But in the absence of orographic controls, fallen snow will melt more quickly in central urban streets and parks than in suburban gardens, where it will often disappear several days before that covering the farmlands around the city.

Because of the higher temperatures and decreased stability in air over cities, hail occurrence as well as heavy rainfall is likely to be enhanced, though the only concrete evidence for this comes from the very special La Porte (U.S.A.) situation described by S. A. Changnon.

Hydrology of cities

Quite apart from any effect cities may have upon precipitation, there are three main physical effects of urban land use upon hydrology. These are changes in total run-off, changes in peak flow characteristics, and changes in water quality.

The two principal controls upon run-off and regime are the percentage of impervious surfaces and the rate at which water is carried across the land. Urban areas will obviously affect both variables. Not all urban fabrics are impervious; indeed, many such as bricks and tiles are highly absorbent. Also, the percentage of the total urban area which is built upon is very variable, both within and between cities. In general, however, run-off from urban areas can be expected to be greater than from orographically similar rural areas.

There have been many independent inquiries, yielding a variety of results. Figures range from a 50 per cent increase in mean annual flood discharge for a one-square-mile area which is 20 per cent impervious and is 20 per cent covered by storm sewerage, to a 400 per cent increase for a one-square-mile area which is 80 per cent impervious and 80 per cent covered by storm sewerage. For unsewered areas, an increase from 0 to 100 per cent imperviousness will increase peak discharge on average 2·5 times; for areas that are 100 per cent sewered, the ratio of discharge after urbanization to that before rises from about 1·7 with zero imperviousness to about 7·0 with imperviousness of 100 per cent.

As run-off is increased by urbanization, so, for the same reasons, soil and ground water storage and replenishment are reduced. In their turn, these will result in decreased low flows. Urbanization therefore has the effect of intensifying the extremes of flow. This means that the recurrence interval in years of a given discharge is also sharply reduced by urbanization.

The influence of urbanization upon the chemical composition of surface and ground water has attracted much attention in recent years as one of the most serious aspects of environmental pollution. The typical city dweller uses daily 150 gallons (700 litres) of water, 4 pounds (2 kg) of food and 19 pounds ($8\frac{1}{2}$ kg) of fossil fuels. This is eventually converted into roughly 120 gallons of sewerage (assuming an 80 per cent recovery of the water input) and 4 pounds of solid waste. Though only occasionally a problem in the developed world, water-borne diseases in polluted water supplies are among the main causes of morbidity and mortality in cities of the under-developed countries.

Visibility

Because of the high aerosol concentrations in urban areas, visibilities are generally lower, and because of the particulates, fog droplets form more readily and evaporate more slowly than in rural areas. The densest fogs are often found in the suburbs rather than in city centres, which are warmer and have lower absolute humidities. Central London, for instance, had, before the Clean Air Act of 1956, nearly twice as many hours of fogs (visibilities of less than 1000 m) per year as rural areas around the city, but only about the same number of hours of dense fog (visibilities of less than 40 m). The London suburbs, on the other hand, had only between 28 and 32 per cent of the fog hours that rural areas beyond the city had, but up to 4 times as many dense fogs.

The warmer air and gentler winds of the city centres also delay the formation of evening fogs and their dispersal the following morning, so that in the evening the fog lies like an annulus around a gradually disappearing clear centre; and in the morning the fog clears in the stronger winds and more rapidly-rising temperatures of rural and suburban areas, to leave fog in only the centre of the city.

Since the Clean Air Acts (1956 and 1968) in Britain, there has been a dramatic improvement in visibilities in London; similar improvements have been recorded in Manchester and elsewhere. However, some of the improvement might be because of stronger winds. It is certainly not universal.

Conclusion

When field, farm and forest are replaced by brick, concrete and macadam, the meteorological properties of the air above, between and within, buildings will be changed. Some of these changes, such as the longer plant-growing season and smaller space-heating loads which result from the higher temperatures in cities, improve the general quality of urban life. Others, such as the higher pollution concentrations found in most cities, are less welcome.

It is important to consider the attributes of urban climates in all considerations of urban ecology and planning.

FURTHER READING

Angell, J. K., Pack, D. H., Dickson, C. R. and Hoeker, W. H. (1971) "Urban influence on night-time airflow estimated from tetroon flights." *Jour. App. Met.*, **10,** 194–204.

Bach, W. (1971), "Atmospheric turbidity and air pollution in Greater Cincinnati." *Geog. Review*, **61,** 573–94.

Bach, W. (1972), "Urban Climate, Air Pollution and Planning" in *Urbanization and Environment* (Ed. Detwyler, T. R. and Marcus, M. G.), Duxbury Press, Belmont, California, 68–96.

Bornstein, R. D. (1968), "Observation of the urban heat-island effect in New York City." *Jour. App. Met.*, **7**, 575–82.

Bornstein, R. D., Lorenz, A. and Johnson, D. S. (1972), "Recent observations on urban effects on winds and temperatures in and around New York." In *Preprints, Conference on Urban Environment and Second Conference on Biometeorology*. American Meteorological Society, Philadelphia, 28–33.

Chandler, T. J. (1965), *The Climate of London*, Hutchinson, London, 292 pp.

Chandler, T. J. (1968), *Selected Bibliography on Urban Climate*, World Meteorological Organization, Pubn. No. 276, T.P. 155, 383 pp.

Changnon, S. A. (1972), "Urban effects on thunderstorm and hailstorm frequencies." In *Preprints, Conference on Urban Climates and Second Conference on Biometeorology*, American Meteorological Society, Philadelphia, 177–85.

De Boer, H. J. (1966), *Attenuation of Solar Radiation due to Air Pollution in Rotterdam and its surroundings*. Koninklijk Nederlands Meteorologisch Institut, Wetenschappelijk Rapport W. R. 66–1, de Bilt, Nederlands, 36 pp.

Dronia, H. (1967), "Der Stadteinfluss auf den weltweiten Temperaturtrend." *Meteorologische Abhandlungen*, **74**, 1.

Graham, I. R. (1968), "An analysis of turbulence statistics at Fort Wayne, Indiana." *Jour. App. Met.*, **7**, 90–3.

Hass, W. A., Hoeker, W. H., Pack, D. H. and Angell, J. K. (1967), "Analysis of low-level, constant volume balloon (tetroon) flights over New York City." *Quart. J. R. Met. Soc.*, **93**, 483–93.

Jenkins, I. (1969), "Increase in averages of sunshine in central London." *Weather*, **24**, 52–4.

Johnson, D. S. and Bornstein, R. D. (1974), "Urban-rural wind velocity differences and their effects on computed pollution concentrations in New York City." In *Preprints of Am. Met. Soc. Symp. on Atmos. Diff. and Air Poll.*, Santa Barbara, California, Sept. 1974.

Lee, D. O. (1975), *A study of the Influence of Surface Roughness and Temperature of Cities on Urban Airflow*. Unpublished Ph.D. Thesis, University of London.

Mateer, C. L. (1961), "Note on the effect of the weekly cycle of air pollution on solar radiation at Toronto." *Int. Jour. Air Water Poll.*, **4**, 52–4.

Maurain, C. H. (1947), *Le Climat Parisien*. Presses Universitaires, Paris, 163 pp.

Meetham, A. R. (1945), *Atmospheric Pollution in Leicester*. D.S.I.R. Technical Paper No. 1, H.M.S.O., London, 161 pp.

Mitchell, J. M. (1953), "On the causes of instrumentally observed secular temperature trends." *Jour. Met.* **10**, 244–61

Nader, J. S. (1967), *Pilot Study of Ultraviolet Radiation in Los Angeles, October 1965*. Public Health Service Publication, 999–AP–38. Nat. Center for Air Poll. Control, Cincinnati, Ohio.

Oke, T. R. (1974)), *Review of Urban Climatology*, 1968–1973, Technical Note No. 134, World Meteorological Organization, Geneva, 132 pp.

Oke, T. R., and Hannell, F. G. (1970), "The form of the urban heat island in Hamilton, Canada." In *Urban Climates*, Technical Note No. 108, World Meteorological Organization, Geneva, 113–26.

Oke, T. R., Maxwell, G. B. and Yap, D. (1972), "Comparison of urban/rural cooling rates at night." In *Preprints, Conference on Urban Climates and Second Conference on Biometeorology*, Amer. Meteorol. Soc., Philadelphia, 17–21.

Roach, W. T. (1961), "Some aircraft observations of fluxes in solar radiation in the atmosphere." *Quart. J. R. Met. Soc.*, **87**, 346–63.

Stair, R. (1966), "The measurement of solar radiation, with principal emphasis on the ultraviolet component." *Int. J. Air Water Poll.*, **10**, 665–88.

United States Environmental Protection Agency (1970), Proceedings of Symposium on Multiple-Source Urban Diffusion Models, APCO Pubn. No. AP–86.

CHAPTER TWO

DESIGN AND BUILDING FOR A TROPICAL ENVIRONMENT

FIKRY N. MORCOS-ASÄAD

Climate and human shelter

Throughout the history of human civilization, shelter against the rigours of the environment has been one of man's basic needs. His inventiveness enabled him to devise a suitable and sometimes elaborate defence against hostile climates. History shows us that with accumulated human experience and imagination, the architecture of the shelter evolved into diversified solutions to meet the challenges of widely varying climates, indicating that the ancients recognized regional adaptation as an essential principle of architecture.

Climate has a pronounced effect on human temperament and physiology. Julian Huxley relates human history to climate by comparing the incidence of early civilization with that of dry and wet epochs, and speculates that the biological and economic effects of the shifts in climatic belts held the balance for populations. Ellsworth Huntington in *Principles of Human Geography* (1951) ranked climate with racial inheritance and cultural development as the three most significant factors affecting conditions of civilization.

Considering the technological limitations of a period and a locale, and the always overriding aspects of safety, the traditional building forms of the rural tropics* show sound solutions of climatic problems. Indeed some of these solutions could be considered ingenious. The temperate areas offered

* Tropical climates are those where heat is the dominant problem and where buildings for the greater part of the year serve to keep the occupants cool rather than warm.

a naturally favourable climate and made few thermal demands on their inhabitants, allowing freedom in structures and a corresponding diversity of form. However, the hot arid areas, characterized by excessive heat and bright sun, made extreme demands on their inhabitants. Shelter was needed to reduce the heat impact and to provide shade. This was achieved by compact planning about closed courtyards, using plan forms with minimum exposed external surfaces. On the other hand, the warm-humid zones presented two problems to constructors: avoidance of excessive solar radiation, and provision for moisture evaporation by breezes. To cope with these problems, the settlements were built to allow free air movement. The roofs were insulated and provided with large overhangs to protect against sun and rain and to provide shade. The floors were raised to keep them dry and to allow for air circulation underneath.

In various regions, people adapted their dwellings to their particular environmental requirements, and an awareness of climate was successfully integrated with local technology, materials and craftsmanship to solve problems of comfort and protection. This resulted in buildings that were

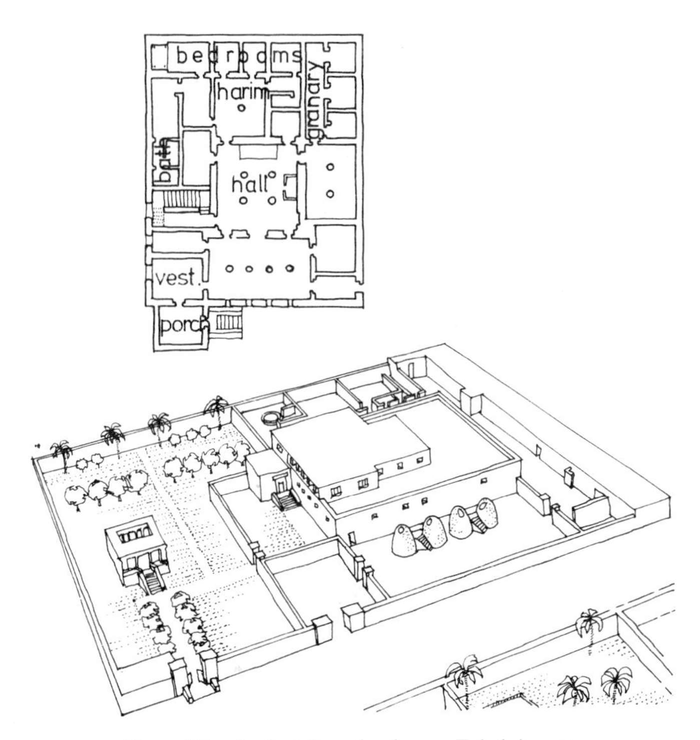

Figure 2.1 Ancient Egyptian house, Tal-el-Amarna

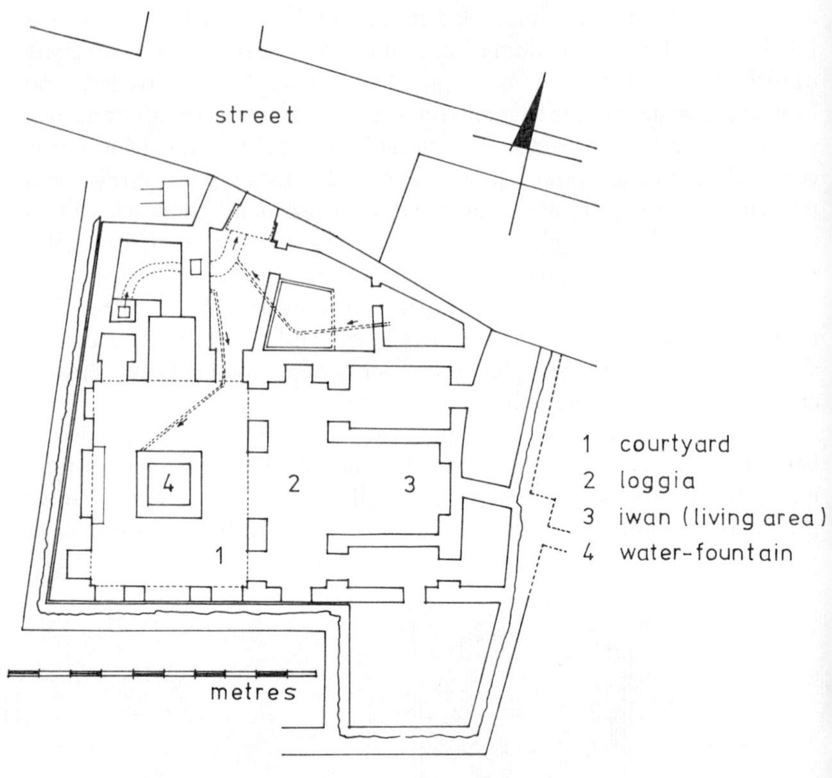

Figure 2.2 House Plan, Fostat, 19th century, Ibn-Touloun period, Egypt

truly expressive of their regional character. Yet today's most pressing requirement in the tropics is housing; and traditional forms, because of their origin in the life and economy of rural societies, are seldom suited to urban conditions. Building materials for town needs can no longer be taken from the surrounding countryside, and materials that have to be brought in over large distances cease to be economic.

Problems of urbanization in the tropics cannot be solved by the adoption of Western technology and patterns that have their origins in different climates, cultures and economic conditions. House types and building materials suitable for cold climates will not solve the problems of towns where heat is the dominant factor. Also, the solutions imported from communities with average *per capita* incomes of £4000 per annum cannot work in communities with average *per capita* incomes of £70. Yet the cities of the tropics are today full of curtain-walled buildings, plate-glass picture windows, and architecture that could be in Glasgow, Chicago or New

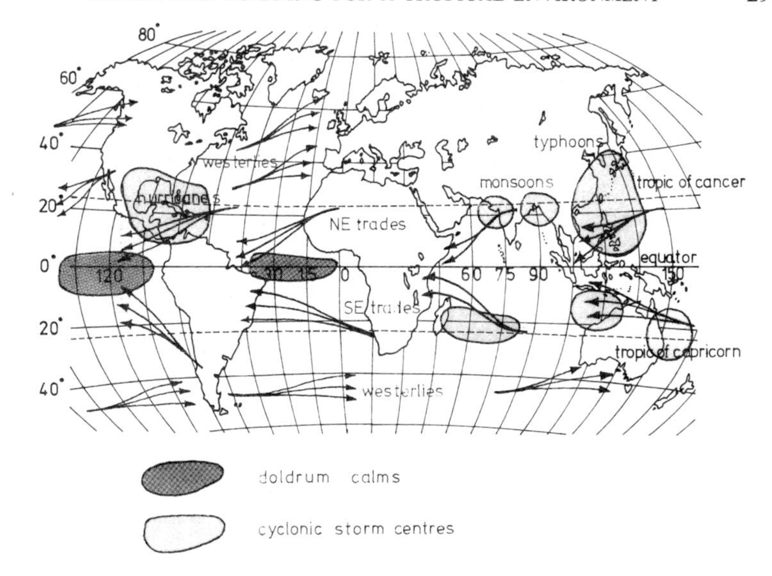

Figure 2.3 Guide to wind systems and air pressure zones

York; and the resulting urban environment is both climatically and socially inadequate.

This situation need not exist. However, it is possible to create the urban setting that will provide pleasant indoor and outdoor living spaces suited to the social and economic conditions of the region's inhabitants only by knowledge of the factors which shape the particular climate, and of the methods used in the design of buildings to provide shelter and comfort in that climate.

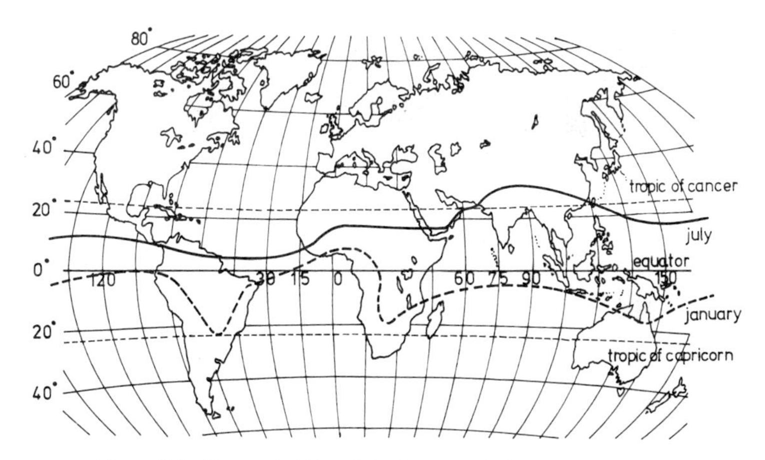

Figure 2.4 Seasonal shifts of the inter-tropical convergence zone

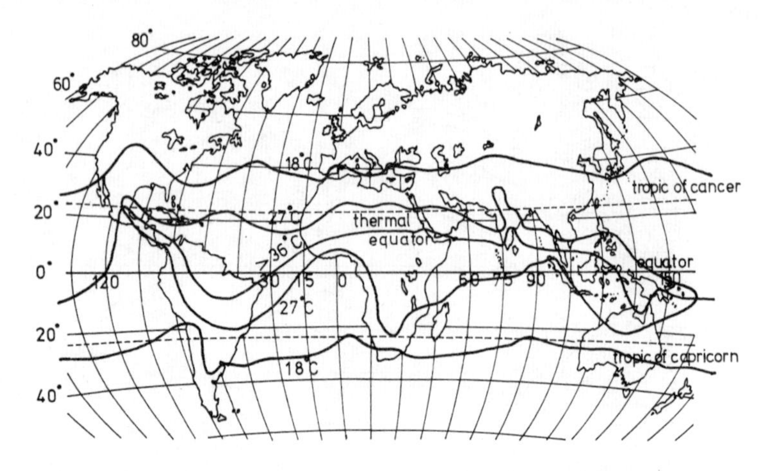

Figure 2.5 Approximate annual temperatures in the tropics

Factors shaping climate

Climate is shaped by solar radiation which warms the earth's atmosphere and ground surface. The earth in turn re-radiates heat to space. This radiant cooling and solar heating create temperature differentials between land masses and oceans, and between the north and south poles and the

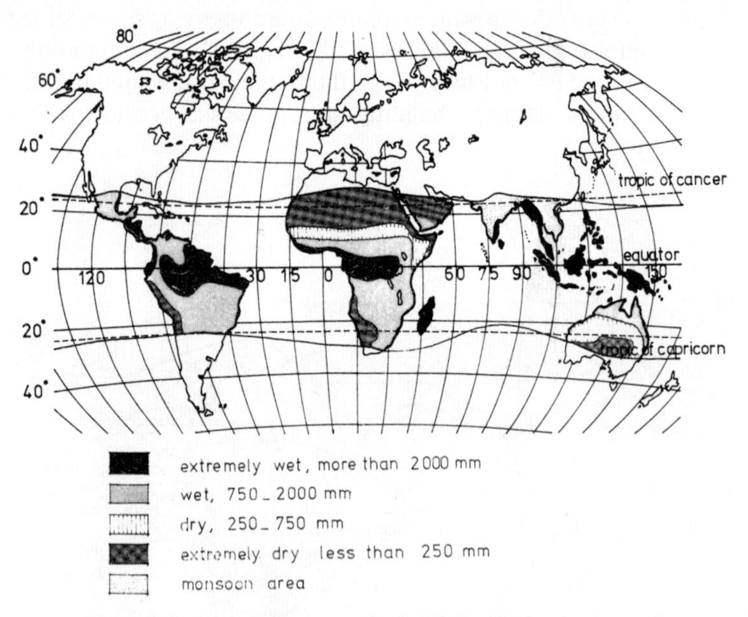

Figure 2.6 Approximate annual rainfall distribution in the tropics

equator. These temperature differentials produce pressure systems which in turn generate winds and air movements (figure 2.3). Moisture from oceans and land surfaces rises before condensing and falling as rain (figure 2.6).

The main climatic elements to be considered when designing for human comfort are:

(*a*) *Air temperature*—representing the earth's thermal balance and measured in degrees C. Air temperaure needs to be lower than the temperature of the skin to cool it by convection.
(*b*) *Air movement*—resulting from thermal forces and wind flow. Wind is measured by its velocity (in m/s at 10 m above ground level), its direction and frequency.
(c) *Solar radiation*—The energy arriving at a surface from the sun in a given time can be measured in watts per square metre (W/m^2).
(*d*) *Humidity*.
(i) The absolute humidity of the air is measured as the amount of moisture present in unit mass of unit volume of air (AH).
(ii) The relative humidity indicates the evaporation potential

$$RH = \frac{\text{absolute humidity at the prevailing temperature}}{\text{maximum amount of moisture the air can hold at the same temperature}} \times 100\%$$

(*e*) *Precipitation*—For all forms of water depositied from the atmosphere (as rain, snow, hail, dew and frost), it is expresed in mm/h, mm/day or mm/month.
(*f*) *Topography and vegetation*—characteristics of site that will influence building design.

Climate and human comfort

The evaluation of human comfort requirements in different types of tropical climates is usually based on the mechanisms of heat exchange between the human body and its environment. This heat exchange process depends on four basic factors: air temperature, humidity, air movement, and radiation. These directly affect the thermal body balance, as each may help or hinder the dissipation of surplus heat from the body.

As an example of the effect of the four climatic variables on the heat dissipation process of the human body, consider an indoor climate of calm ($<0{\cdot}25$ m/s) warm (18°C) air, with moderate humidity (40–60%), where the temperature of the bounding surface is the same as that of the air; for sedentary work the surplus heat dissipated by the body will be 45% by radiation, 30% by convection, and 25% by evaporation.

Thermal body balance and heat exchange

The deep body temperature must remain at a balanced and constant 37°C and, in order to maintain this, the body should be able to release all its surplus heat to its environment by convection, radiation, evaporation, and to a lesser degree by conduction.

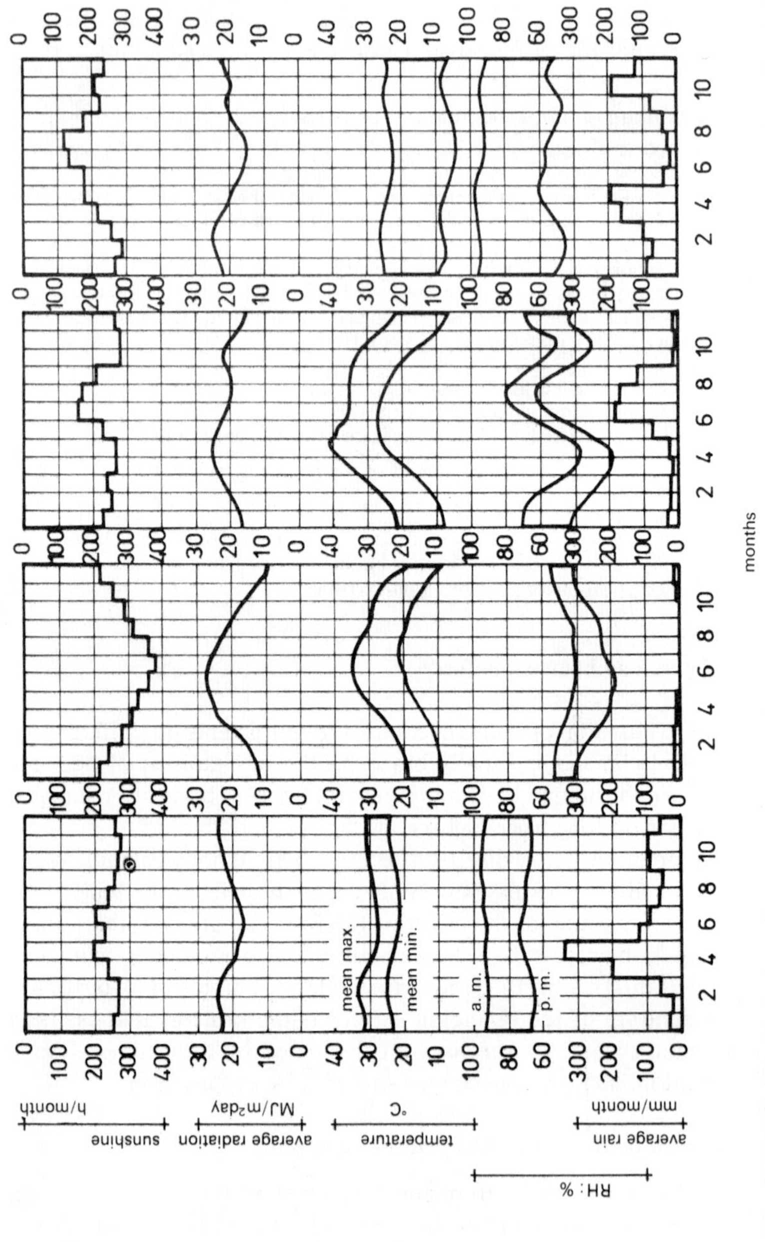

Figure 2.7 Climate graphs for cities in different time zones

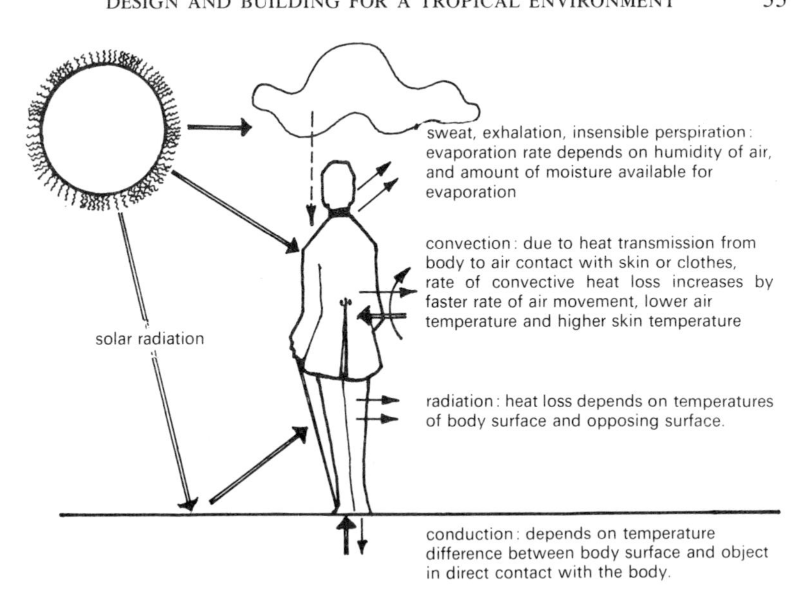

Figure 2.8 Thermal body balance and heat exchange

This thermal balance of the body may be expressed as

$$M - E \pm C_d \pm C_v \pm R = 0$$

where M = metabolism. Of all the energy produced in the body, only about 20% is utilized, and the surplus 80% must be dissipated as heat. The rate of excess heat output varies with activity, from about 50 watts when sleeping, 150 watts for moderate movement, 300–400 watts for walking and moderate lifting, up to 600–800 watts for sustained hard work.

E = evaporation of moisture and sweat. As the latent heat of water is 2400 kJ/kg, then the evaporation rate of 1 kg/h will produce a heat loss rate of 2 400 000 ÷ 3600 = 667 watts.

C_d = conductive heat exchange—negative if the body is in contact with cold bodies, positive if the body is in contact with warm bodies.

C_v = convective heat exchange—negative if the air is cooler than the skin, positive if the air is warmer than the skin.

R = radiant heat exchange—negative to night sky and cold surfaces, positive from the sun and hot bodies.

If the result of the above summation is more than zero, vasomotor adjustments will take place, and blood circulation to the skin surface will increase to accelerate body heat loss processes. Conversely, if the resulting sum is less than zero, the blood circulation to the skin is reduced, and heat loss processes are slowed down. If the vasomotor adjustment is still insufficient and overheating continues, sweating will start. The rate of sweating, during periods of physical effort combined with hot environmental effects, may vary from 20 g/h to 3 kg/h (loss rate of 14 to 2000 watts).

As might be expected, inhabitants of warmer climates prefer somewhat higher temperatures than those living in cooler regions. Acclimatization processes involving an increase in the quantity of blood to produce and maintain a constant vaso-dilation and an increase in sweat rate will result from long-term exposure of the human body to a new set of warm climatic conditions. Full adjustment is reached in about thirty days.

Thermal comfort and discomfort

Physiologically, comfort is a condition under which the thermo-regulatory mechanisms of the body are in a state of minimal activity. Thermal comfort is a subjective assessment of the environment and should not be confused with equilibrium in the heat exchange of the human body with the environment. This equilibrium or heat balance, while essential for comfort, may also be achieved under conditions of discomfort through the activation of the previously mentioned thermoregulatory vasomotor mechanisms.

Maintaining thermal comfort does not imply the maintenance of the indoor thermal conditions at a constant level, as the vasomotor thermoregulatory systems of the body are capable of providing physical comfort within a given zone of conditions. In fact, some slight fluctuations in the indoor conditions, such as in air velocity and temperature, can be invigorating and beneficial in preventing a feeling of monotony. The indoor thermal requirements of the inhabitants may be specified within a zone where some variations and fluctuations are accepted. This zone is known as the *comfort zone*.

Beyond this zone, a sensation of discomfort is experienced as a result of the deviation of one or more functions of the environmental factors from their level at comfort conditions. In hot climates, two kinds of thermal discomfort can be distinguished.

(*a*) *Thermal sensation of hotness.* This is closely associated with dry heat exchange with the environment by convection and radiation. Under constant activity conditions it is linearly correlated up to certain limits with the physiological responses of skin temperature in cold exposure, and with sweat rate under hot exposure.

(*b*) *Wetness of the skin in the form of sensible perspiration.* This is experienced only on the warm side of the comfort zone and in specific combinations of temperature, humidity, air velocity and metabolic rate.

These two sensations are affected differently by the various environmental factors. Both types may be experienced simultaneously but, under other conditions, one may be felt without the other. For instance, air movement affects body cooling, though it does not decrease the air temperature. The cooling sensation is due to increased evaporation from the body and to heat loss by convection. As the air velocity increases, the upper comfort level is raised. However, this rise slows as higher temperatures are reached.

Comfort zones

Ideas of what is comfortable vary from one person to another, but research has established remarkable agreement on the upper and lower limits beyond which at least 70% of test subjects complained of discomfort. Comfort zones were established for a number of locations, expressed in air temperatures which vary regionally according to humidity and annual mean temperature.

An approximate guide to comfort limits according to annual mean temperatures and relative humidity is shown in Table 2.1.

Table 2.1 Comfort limits given in *Climate and House Design*, Vol. 1., United Nations, 1971.

Annual mean temperature	*over* 20°C		15–20°C		*under* 15°C	
Average relative humidity	Day	Night	Day	Night	Day	Night
0–30%	26–34°C	17–25°C	23–32°C	14–23°C	21–30°C	12–21°C
30–50%	25–31°C	17–24°C	22–30°C	14–22°C	20–27°C	12–20°C
50–70%	23–29°C	17–23°C	21–28°C	14–21°C	19–26°C	12–19°C
70–100%	22–27°C	17–21°C	20–25°C	14–20°C	18–24°C	12–18°C

However, the use of one of the following *comfort scales* for translating the regional and site climatic data, and as a guide in establishing thermal comfort criteria for the indoor climate, is recommended.

Effective Temperature Scale (ET) This scale, produced by Houghton and Yaglou and later modified, integrates the effects of three variables (temperature, humidity, and air movement) in the form of "equal comfort lines" (figure 2.9). The scale underestimates the significance of moderate air movements at high temperatures and overestimates the adverse effect of higher humidities.

Corrected Effective Temperature (CET) This is similar to the ET scale, with appropriate modifications for air movement. It also includes radiation effects. Although its accuracy is questioned by some researchers, it is the most widely used scale at present. However, it still does not allow for radiation heat exchange between the body and its environment.

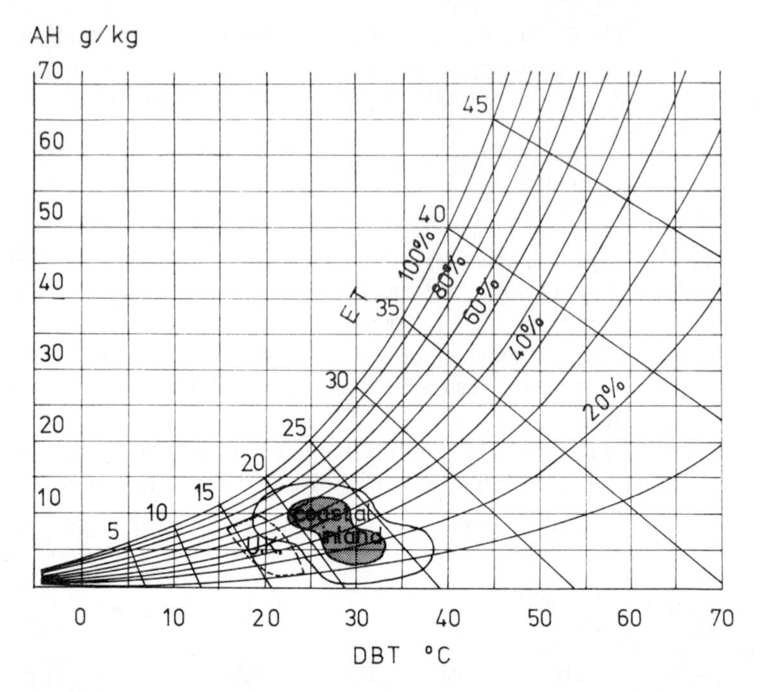

Figure 2.9 Psychrometric chart with effective temperature lines (after Koenigsberger)

Figure 2.10 shows nomograms (*a*) and (*b*) of the CET scales for persons dressed in lightweight normal clothing and for persons stripped to the waist. The CET for a given situation is obtained by the following procedure:

1. Measure the globe temperature (GT) or the dry-bulb temperature (DBT). These are synonomous, and represent the true air temperature.
2. Measure the wet-bulb temperature (WBT).
3. Measure the air velocity.
4. Locate the GT and WBT on the left- and right-hand scales.
5. Connect the two points.
6. Select the curve appropriate to the air velocity, and mark where the velocity curve intersects the line drawn in (5).
7. Read off the value of the short inclined line throught the same point. This is the CET value.

The Bioclimatic Chart was drawn up by Olgyay to define a comfort zone in terms of dry-bulb temperature and relative humidity. The additional lines show how this comfort zone can be moved upwards in the presence of air movement and lowered by radiation (figure 2.11).

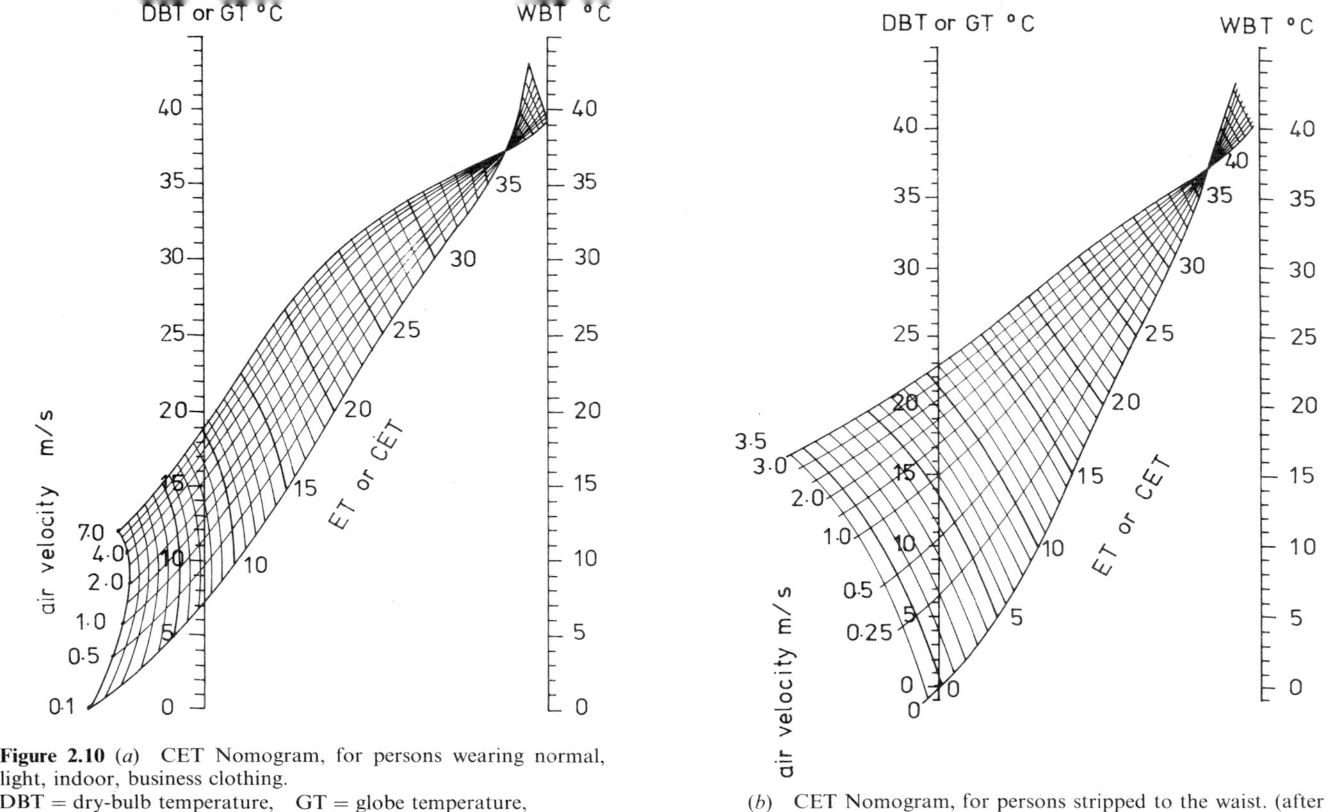

Figure 2.10 (*a*) CET Nomogram, for persons wearing normal, light, indoor, business clothing.
DBT = dry-bulb temperature, GT = globe temperature,
WBT = wet-bulb temperature.

(*b*) CET Nomogram, for persons stripped to the waist. (after Koenigsberger)

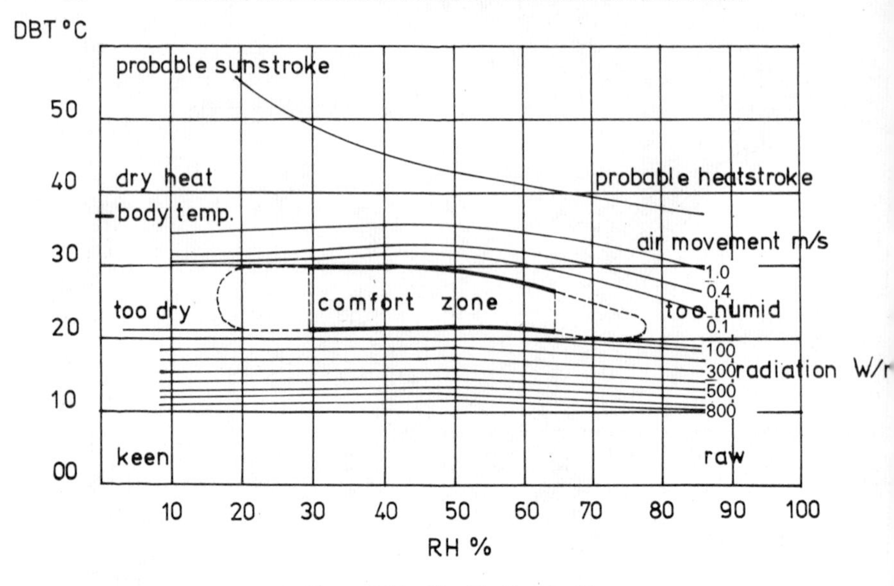

Figure 2.11 The Bioclimatic Chart

Design and climatic control

In the design of buildings appropriate to a given climate, three distinct aspects have to be considered:

(*a*) the climate of a specific region
(*b*) the user's preferred indoor climatic conditions
(*c*) the technical and structural methods used to achieve those indoor conditions (p.41).

The basic requirements from a building are to protect the interior from extremes of solar radiation; to provide for reasonable air movement to assist cooling by evaporation, and to protect the eyes from strain caused by glare.

The range of possible control of heat load operating on the building and the resulting variations in the indoor climate are summarized in Table 2.2.

Table 2.2 Heat load and indoor climate

Climatic variable	*Range of variation in indoor climate*
Solar radiation absorbed in walls	15 to 90% of incident radiation
Solar radiation penetrating through walls	10 to 90% of incident radiation
Indoor air temperature amplitude	10 to 150% of outdoor amplitude
Indoor maximum air temperature	−10 to +10°C from outdoor maximum
Indoor minimum air temperature	0 to +7°C from outdoor minimum
Indoor surface temperature	−8 to +30°C from outdoor maximum
Average internal air speed (windows open)	15 to 60% of outdoor wind speed
Actual air speed at any point in room	10 to 120% of outdoor wind speed
Indoor vapour pressure	0 to 1000 N/m² above outdoor level

However, a design feature which could control one variable of the indoor climate often reduces the possibility of control of the other factors. For instance, intensive ventilation provides high internal air velocities but almost eliminates the possibility of controlling the indoor temperature. The relative benefit of ventilation against temperature control from the aspect of human comfort depends on the type of climate and on the building design details, which will in turn relate to the building use. In general, the relation of indoor to outdoor conditions varies widely with the different characteristics of the building design and construction.

The building and its climate

As mentioned earlier, one of the most important functions of shelter is to provide man with protection against adverse climatic elements which may be injurious to him. The building shell itself will provide the means of thermal control of indoor climate through its siting, its design details, and the materials used in its construction. Thus, in a building the objective in terms of thermal controls may be stated as follows:

When hot conditions prevail:

(*a*) to prevent heat gain
(*b*) to maximize heat loss
(*c*) to remove any excess heat by mechanical cooling.

When conditions vary between hot and cold discomfort:

(*a*) to even out the diurnal temperature variations
(*b*) to prevent heat loss and utilize solar heat gain in the cold period, and to prevent heat gain and maximize heat loss for the hot period
(*c*) to compensate for both cold and hot excesses by a flexible mechanical heating/cooling system

The objectives stated under (*a*) and (*b*) in both cases can be achieved by structural and constructional passive means, while objective (*c*) in each case can be achieved only by an active energy input or by mechanical means of control (figure 2.12).

Thermal balance of building and its heat exchange

A building (figure 2.13) may be considered as a defined unit, with its heat exchange process with the outdoor environment expressed by

$$I - E + S \pm C \pm V \pm M = 0$$

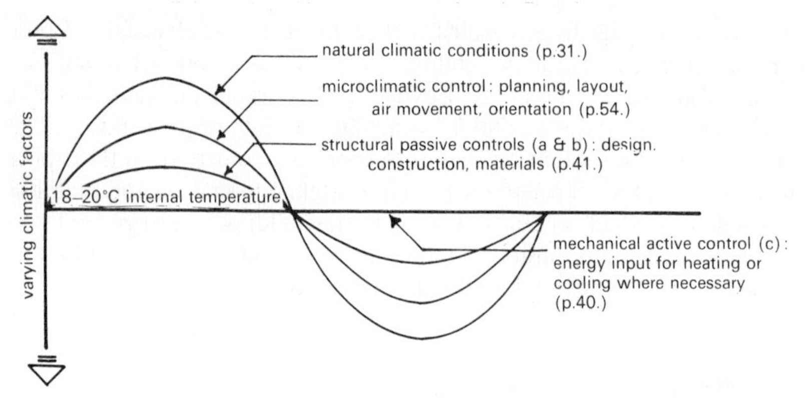

Figure 2.12 Potential of climatic controls

where $I =$ internal heat gain from heat output of occupants, lights and other appliances,
$E =$ evaporation from surface of buildings (roof pools) or from within the building (human sweat, water fountains), and where the vapours are removed to produce a cooling effect,
$S =$ solar heat gain from solar radiation on opaque surfaces and through windows,
$C =$ conductive and convective heat from walls and roof in each direction,
$V =$ ventilation heat exchange from air movement in either direction,
$M =$ mechanically controlled heating or cooling energy input.

This equation indicates that the building's thermal condition is maintained in balance.

If the summation is less than zero, then the building will be cooling down; if the result is more than zero, the temperature of the building will increase.

Mechanical climatic controls

In tropical climates, the need for heating will rarely arise. Probably the tropical upland climates are the only climates where cool, discomfort conditions

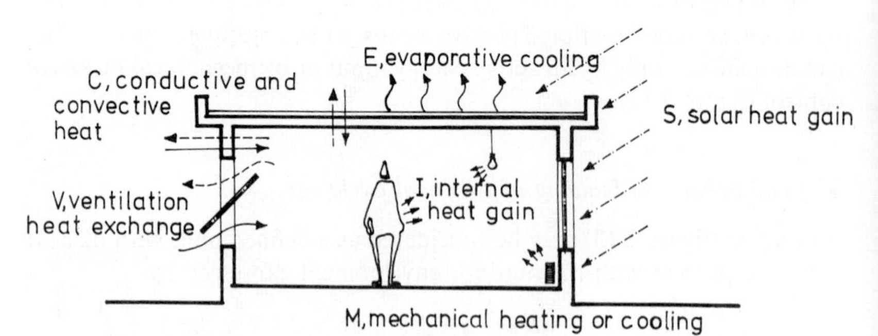

Figure 2.13 Thermal balance of building and its heat exchange

may prevail for long enough to make the thermal storage capacity of the structure insufficient to ensure indoor human comfort. However, as the deficit is small, the incidental internal heat gain from occupants and appliances in a suitably designed and constructed building could increase the indoor air temperature to a level acceptable with warm clothing. At most, some local heating appliances could be provided. In a warm climate it may also be an advantage to provide a slow revolving fan of large diameter to improve the ventilation when there is little natural air movement and excess heat conditons.

Structural climatic controls

As mentioned above, structural passive controls can be effective in providing favourable indoor climatic conditions through natural means. This method of control is usually adequate by itself, to ensure the best possible indoor thermal conditions without the aid of any mechanical active controls. If, however, mechanical means have to be restored to, their task will be reduced to a minimum.

The structural control of the thermal indoor environment is achieved through:

(*a*) the selection of materials for protecting the interior from the extreme effect of solar radiation,
(*b*) the surface characteristics of building materials,
(*c*) the reduction of solar heat gain through windows by shading,
(*d*) the application of thermal insulation,
(*e*) the correct roof design and construction.

(*a*) *Selection of Construction Materials* The correct choice of construction materials in tropical climates is an extremely important aspect of total design. The thermal control characteristic of a material is its transmission of heat behaviour. The daily external heat load variation will cause a corresponding fluctuation inside the building, but with two significant differences:

(i) The inside cycle will be decreased in amplitude, due to the insulation value of the material (known as *U*, the overall heat transfer coefficient). The lower the *U*-value, the better the insulating effect for reducing the heat flow.
(ii) The inside cycle will lag in time behind the outside cycle, i.e. shift in phase, due to the heat-storage value of the material (known as *C*, the heat capacity of a building element). The greater the heat capacity of the building material, the slower the temperature change that is propagated through it. This delay is called the *time-lag* of the construction, which is the difference in time between the occurrence of the maximum external surface temperature and the maximum internal surface temperature under periodic heat flow.

These two characteristics of a material are often expressed together by C/U. The effect of insufficient thermal resistance may be alleviated by an increase in the thermal capacity, and vice versa. However, as the mechanisms of heat flow control operating through the two factors are different, the effectiveness and the relative importance of each in relation to the physiological comfort of the occupants within a building varies with climate type.

Generally speaking, thermal resistance is of greater importance in warm-humid areas where diurnal range is small, while in hot-dry climates with a wide temperature range the heat capacity becomes more significant.

Table 2.3 *U* and time-lag characteristics for some traditional wall materials

Material	*Thickness* (mm)	*U-value* (W/m² °C)	*Time-lag* (hours)
Stone	200	3·81	5·5
	300	3·12	8·0
	400	2·67	10·5
Concrete	100	4·77	2·5
	150	4·22	3·8
	200	3·75	5·1
	300	3·10	7·8
Red brick	100	3·41	2·3
	210	2·33	5·5
	315	1·76	8·5
Timber boards	12	3·86	0·17
	25	2·73	0·45
	50	1·70	1·3

In building design the appropriate utilization of both insulation and the time-lag characteristics of a material are important for the interior heat balance. This heat balance can have a profound effect on the way the inhabitants live within their space. For instance, the villagers in rural Egypt (hot-dry climate) sleep on the roof during the summer months for at least the first half of the night. After midnight they usually move indoors, as by then the outdoor temperature has dropped to a cool discomfort level, while the indoor temperature has become more comfortable.

The value of material insulation becomes more significant as climatic conditions get cooler and design temperatures diverge from comfort requirement. The building mass (weight of material) required for the desirable time-lag varies with the daily temperature fluctuation (diurnal temperature), with the greatest range occurring in hot-dry areas.

(*b*) *Surface Characteristics of Building Materials* The thermal forces acting on the outside of a building are a combination of radiation and convection impacts. The radiation component consists of incident solar

Table 2.4 Reflectivity and thermal emissivity characteristics for solar and thermal radiation of some typical surface materials

	Reflectivity %		*Emissivity %*
Surface	*Solar radiation*	*Thermal radiation*	*Thermal radiation*
Polished aluminium	85	92	8
Whitewash	80	—	—
Chromium plate	72	80	20
White lead paint	71	11	89
White marble	54	5	95
Aluminium paint	45	45	55
Limestone	43	5	95
Wood (pine)	40	5	95
Asbestos cement (1 year old)	29	5	95
Red brick	23–30	6	94
Grey paint	25	5	95
Galvanized iron (aged)	10	72	28
Black matt paint	3	5	95

radiation and of radiant heat exchange with the outside surroundings. The convective heat impact is a function of exchange with the external air and may be accelerated by air movement. The exchange effect may be increased by diluting the radiation over a larger area by the use of curved surfaces such as vaults, domes or corrugated roofs, which will at the same time increase the rate of convection transfer. Another very effective protection against radiation impact is the selective absorptivity* and emissivity** characteristic of a material, especially under hot conditions. Materials which reflect rather than absorb radiation, and which release the absorbed heat as thermal radiation, more readily bring about lower temperatures within the building.

As solar radiation consists of visible and short infra-red radiation, and since this energy is concentrated near the visible part of the spectrum, the criterion of reflectivity* is an approximate relation to colour values. White materials can reflect 90% or more, while black materials may reflect 15% or less of radiation received.

* Absorptivity (a) of a surface is

$$\frac{\text{amount of heat absorbed by surface}}{\text{total amount of heat incident on surface}}\,\%$$

Accordingly, the reflectivity (r) for an opaque surface is the component that is reflected, given by $(1-a)$.

** Emissivity (e) of a surface is

$$\frac{\text{amount of energy emitted by the surface}}{\text{amount of energy emitted by a black body at same temperature}}\,\%$$

On the other hand, the thermal exchange with the surroundings consists of longer infra-red wavelengths, and a material's reflectivity characteristic for long-wave infra-red heat depends more on surface density and molecular composition than colour.

This highly selective behaviour of materials under solar and thermal radiation can be exploited in building construction for various climatic circumstances. If surfaces exposed to solar radiation and to clear sky (as in hot-dry areas) are whitewashed, they will remain cooler than surfaces of polished aluminium, despite the fact that the aluminium surface has a higher reflectivity to solar radiation. This is due to the emissive capability of the whitewashed surface, which loses heat by thermal radiation to the sky. This quality of whitewash accounts for the common white exteriors of tropical buildings. The traditional Egyptian farmer's mud-brick house is lime-washed annually before the summer.

Generally, in zones where overheated periods alternate with cool periods, both reflectivity and absorptivity are desirable at day times. In zones where hot conditions prevail, the net effect of reflectivity combined with the emissive thermal radiation characteristic of a material has to be considered carefully.

(*c*) *Reduction of Solar Heat Gain through Windows* The selection of material has a great significance, as we have just seen. Lesser absorbancy and greater surface conductance reduce the solar heating effect. However, the greatest source of heat gain can be the solar radiation entering through a window, which even in moderate climates, can increase the indoor temperature to far above the outdoor temperature. The direct radiation transmitted varies markedly with the time of day and the angle of incidence, remaining fairly steady until about 50° and dropping sharply after 60°.

Window glasses are practically transparent for short-wave infra-red radiation emitted by the sun, and almost opaque for long-wave radiation emitted by objects in the room. Thus, once the radiant heat has entered through a window, it is trapped within the interior. When solar overheating is a problem, as in all tropical climates, one or more of the following available methods for reducing solar heat gain through windows can be used.

Orientation

In an equatorial location, if solar heat gain is to be avoided, the main windows should face north and/or south. At higher latitudes, an orientation away from the equator receives the least sunshine, but some solar heat may be desirable in winter. In all locations, only minor openings of secondary rooms may be placed on the east and west sides. Solar heat gain on the west side of a building can be troublesome, as its maximum intensity coincides with the hottest period of the day. If all other design and climatic factors are equal, these considerations are valid. However, if wind is to be captured or a pleasant view is to be enjoyed, this desire may override the orientation and solar considerations which can be resolved by other means, such as the use of special glasses or shading devices.

Internal blinds and curtains

The use of internal blinds and curtains on windows is not a very effective way of solar control. It is true that they stop the passage of radiation, but they themselves absorb the solar heat and can reach a high temperature. An internal Venetian blind may reduce the daily average solar gain factor of a single glazed window by up to 17%.

Special heat-absorbing glasses

There are currently a number of heat-absorbing glass types on the market, where the selective transmittance of the ordinary window-glass is modified by varying the material composition. These specially-modified glasses can reduce the infra-red transmission substantially, while affecting the light transmission only slightly. However, one disadvantage is that the reduction in transmittance is accompanied by a corresponding increase in absorbance, so that the glass itself can reach a high temperature. This absorbed heat will then be re-radiated and convected both to the interior and exterior spaces. Thus, the net improvement is not as great as the reduction in transmittance value.

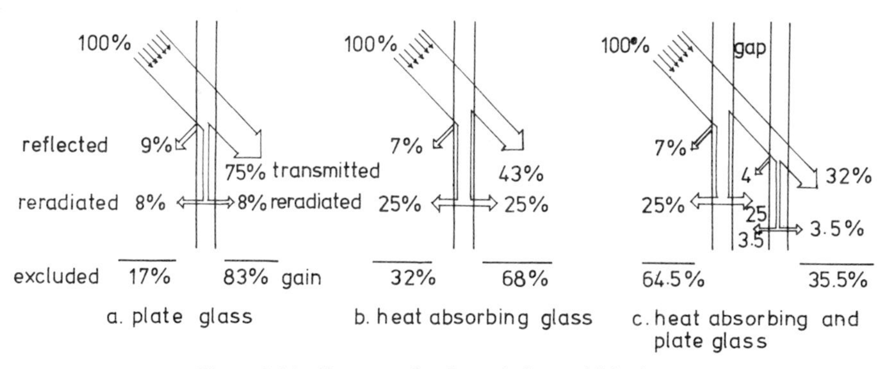

Figure 2.14 Heat transfer through 6 mm thick glasses

One way of overcoming the problem of heat absorption in special glasses is shown in figure 2.14 (*c*) where the panel of heat-absorbing glass is mounted in front of the ordinary plate-glass window at a distance of 0.5–1 m. This will reduce the transmission and, at the same time, the absorbed heat is dissipated on both faces to the outside air. The heat re-radiated to the ordinary glass is at long wavelength, for which it is opaque; thus the total gain outdoors is drastically reduced.

External shading devices (figures 2.15–2.19)

These can be fixed-structural, openable or folding types.

(i) Vertical shading louvered blades or projecting fins—effective when the sun is to one side of the elevation, i.e. east- and west-facing walls.

(ii) Horizontal shading by canopies, horizontal louvered blades, awnings, externally applied Venetian blinds or wooden shutters—effective when the sun is facing the building face and at a high angle, i.e. north- and south-facing walls.

(iii) Egg-crate shading, a combination of horizontal and vertical elements available in many types of grill blocks and decorative screens—effective for any orientation depending on detailed dimensioning.

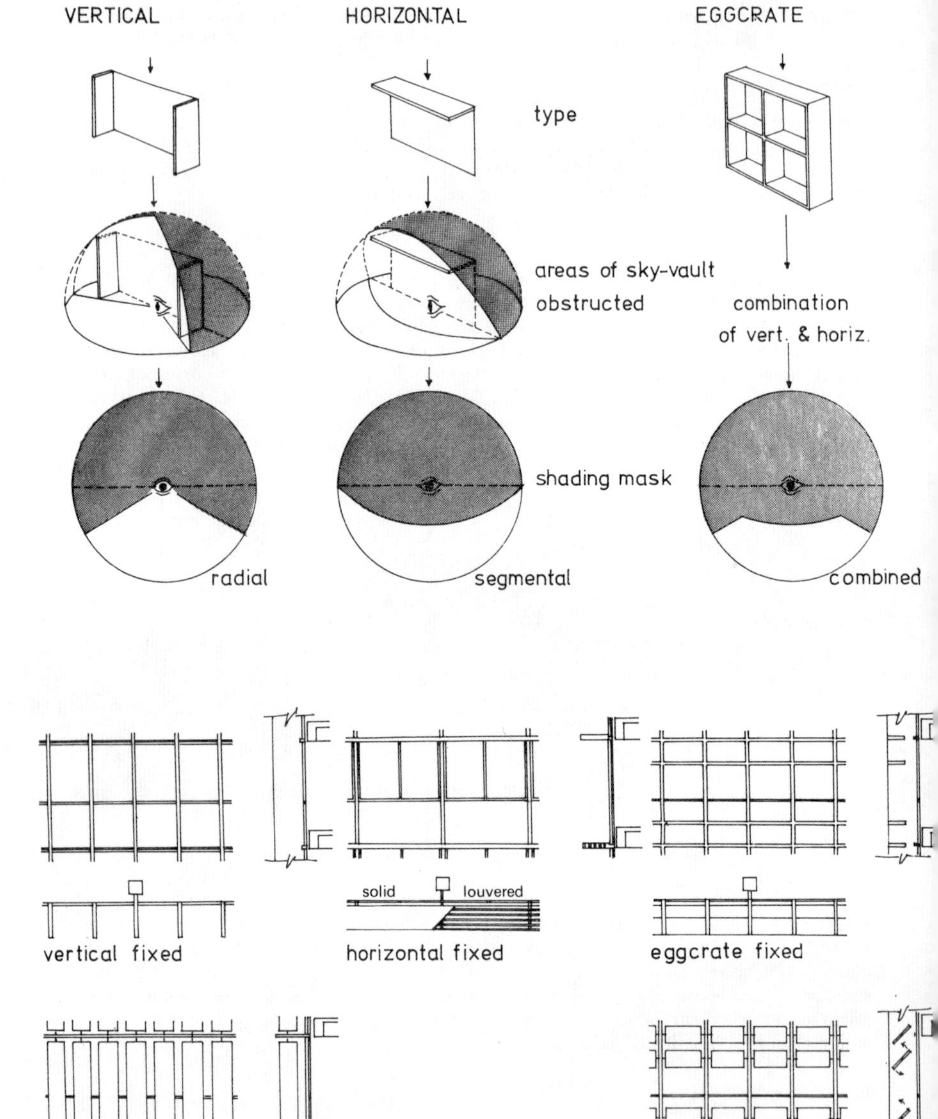

Figure 2.15 Types of shading fins and their masks (after Olgyay)

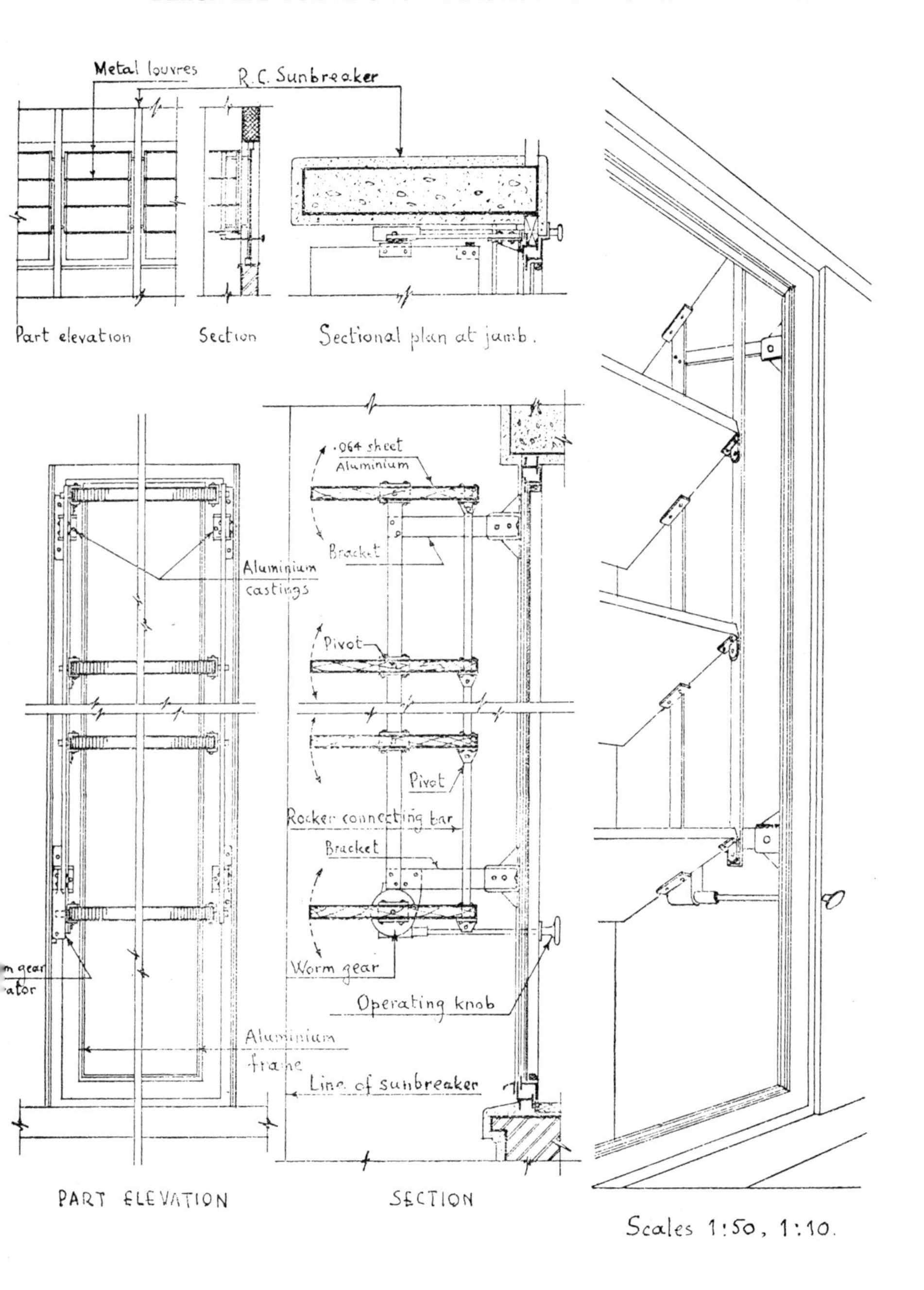

Figure 2.16 Moveable louvers

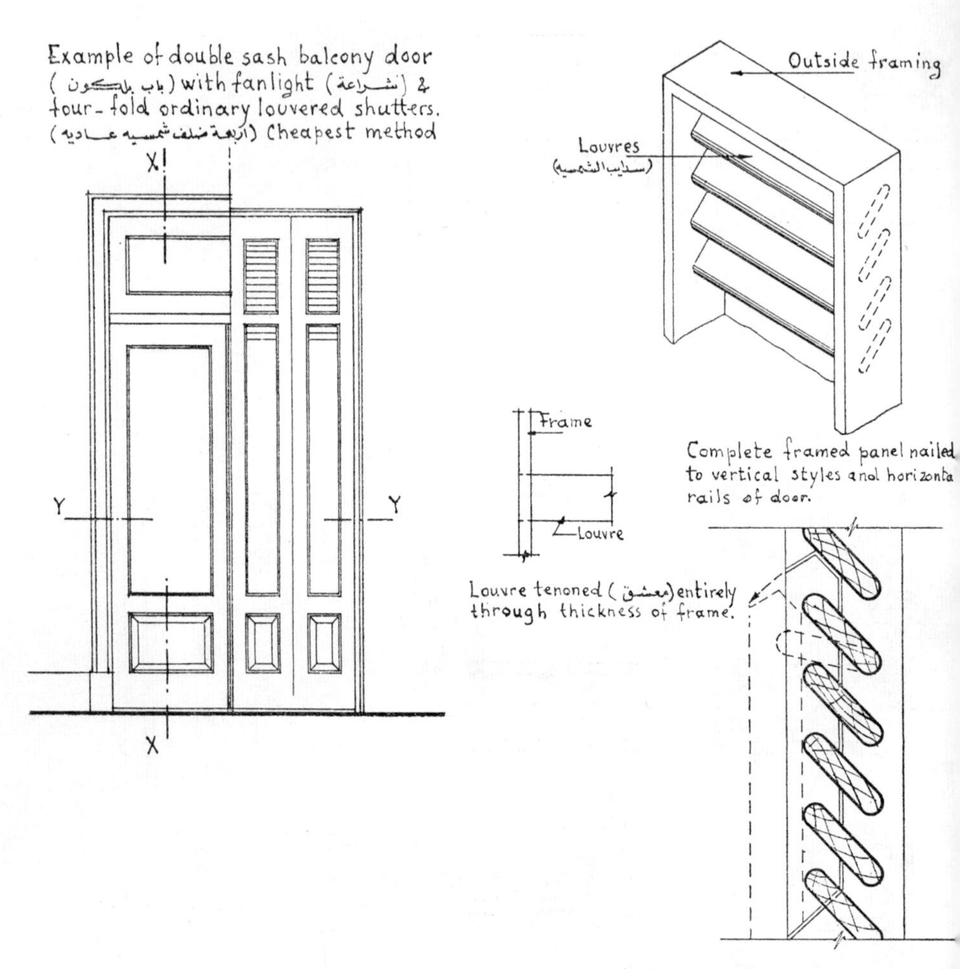

Figure 2.17 Ordinary louvered shutters

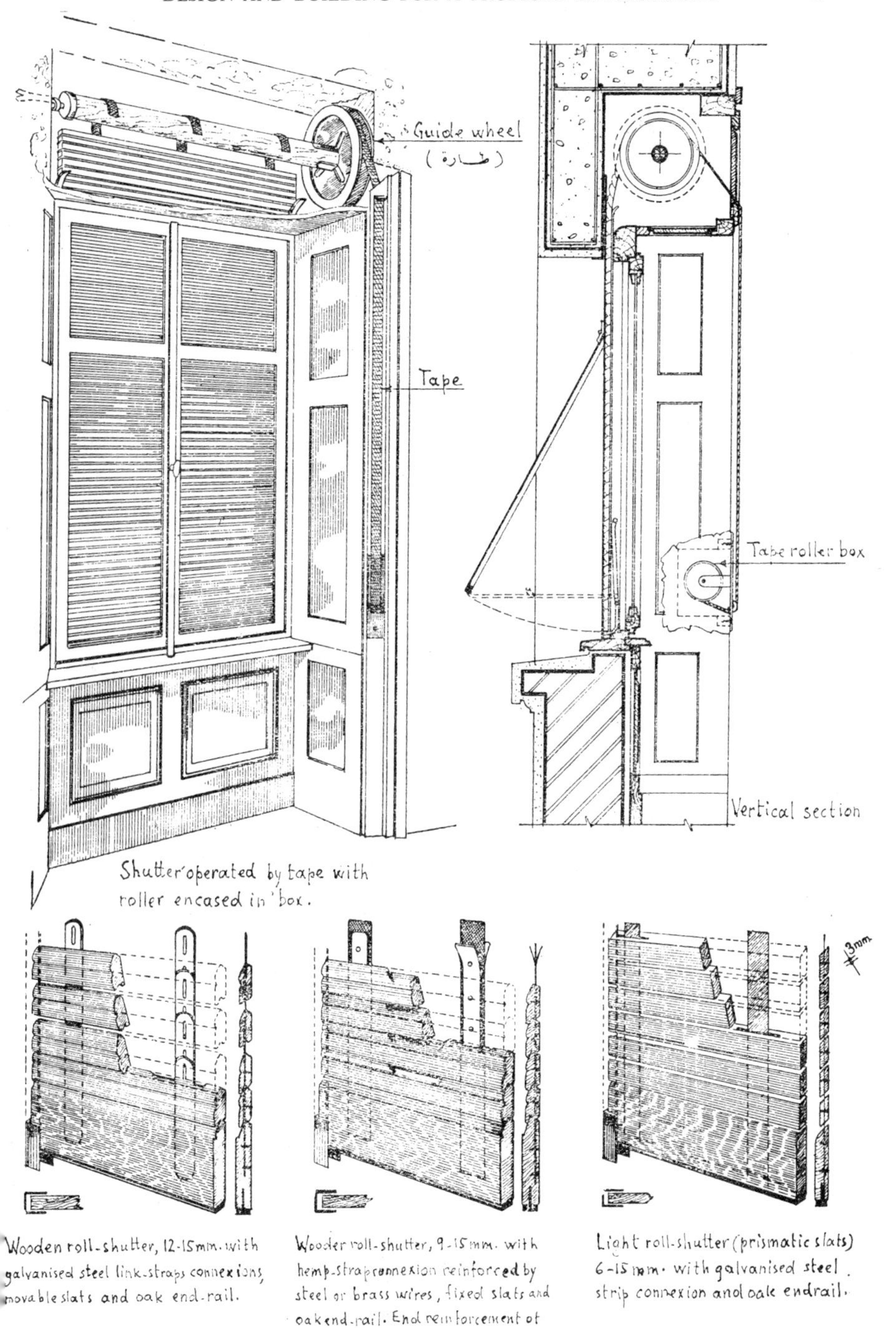

Figure 2.18 Timber rolling shutters

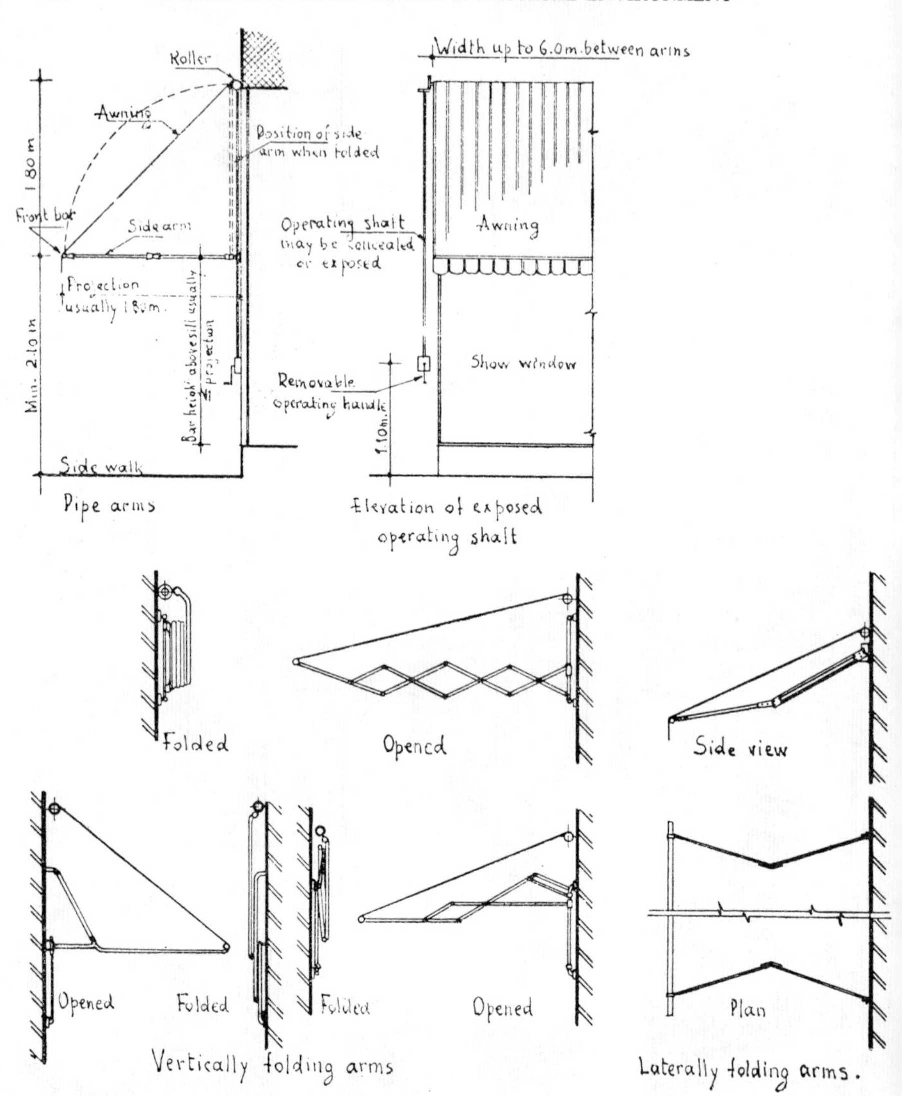

Figure 2.19 Awnings

Shading by trees and vegetation

Trees and vegetation contribute much to the immediate physical environment. They reduce airborne sounds if densely planted; the viscous surface of their leaves catches the dust and filters the air; they can secure visual privacy; and they reduce annoying glare effects. An especially beneficial effect of trees is their thermal performance. The surfaces of grass and leaves absorb radiation, and their evaporation processes cool the air. Trees also provide generous shade at the right seasons. This makes deciduous trees especially valuable when placed close to buildings.

Vines are another of nature's automatic heat-control devices, cooling by evaporation and providing shade. This combination makes them valuable in hot weather.

Climate and soil conditions permitting, both vines and trees should be selected for their appearance, as well as for their shading performance. It is desirable to transplant shade trees in as large a size as practicable; for a tree to give results in a comparatively few years, it should be 5–7 m high when planted. Some fast-growing five-year trees, when planted in new locations, would only take another five years to grow to 80% of their full shading effect.

For trees to give their best shading performance, they should be placed strategically on east-south-east and west-south-west locations from a building, as the sun passes in the morning and late afternoon at a low altitude.

(*d*) *Applied Thermal Insulation* Various insulating materials are available in sheet, loose fibre or foam forms, for application in building construction. They are most effective under steady-state conditions, or at least when the direction of the heat flow is constant for long periods of time, as in heated or air-conditioned buildings. Where the direction of heat flow is twice reversed in every 24-hour cycle, the significance of insulation is diminished.

Heat insulation may be applied either

(*a*) on the outside, where it reduces the heat flow rate into the building mass, or
(*b*) on the inside, where it reduces heat emission to the inside space, but will not change the periodicity. In hot climates the requirement is to store during the day as much as possible of the heat that has entered the outer surface, as well as to dissipate most if not all of this stored heat during the night. Thus by the following morning the heat store of the structure is empty, or contains as little as possible, ready for another storing cycle.

Applied insulation layers on the inside will restrict not only the entry of heat but also its dissipation. If insulation is applied on the outside, which is preferred, then the heat stored is only dissipated effectively to the inside space. To remove this heat, good ventilation for the inner surface by cool night air is necessary.

Another method of insulation commonly used in wall construction is the unventilated cavity system, which is a good insulator. Following the recommendation that insulation should be on the outside of the main mass, then the inner leaf of a cavity wall is considered as the main mass, while the outer leaf is built of lightweight construction.

(*e*) *Type of Roof Construction and its Effect on Indoor Climate* The roof is the building component most exposed to climatic elements. The impact of solar radiation on clear days in summer, loss of heat by long-wave radiation during the night and winter, and rain, all affect the roof more than any other part of a structure. Under warm ambient conditions, the indoor temperature is affected by the roof to varying degrees depending on certain details. In hot countries it is popularly believed that the roof is the main heating element of a house. This is so in the majority of cases, but only because such roofs are incorrectly designed.

The external surface of the roof is often subject to the largest temperature fluctuations, depending on its type and colour, which determine the amount of solar radiation absorbed during the day, the amount of long-wave radiative heat loss into space at night, and consequently the pattern of external surface temperatures and internal heat exchange.

The most common types of roofs are flat and pitched. In the main, the choice depends on functional and economic considerations. However, other design aspects such as aesthetic, shading, and visual and cultural factors, may influence the final solution.

(*a*) *Flat roofs*

Flat roofs are usually of reinforced-concrete slab-type construction and, when uninsulated, they are unlikely to provide adequate protection for residental buildings from tropical heat. The maximum ceiling temperature in the summer season is usually too high and the time-lag is too short, typically 2·5 hours for a 100 mm concrete section (see Table 2.3).

From various researches, it has been found that the temperatures of ceilings under grey roofs are higher than those of the upper air layer, indicating a heat flow from the roof into the house. On the other hand, ceiling temperatures in houses with whitewashed roofs are below those of the upper air layer and the room air during most of the day, implying a flow of heat from the room into the roof. In this instance, the roof acts as a cooling element for the building, because the average external surface temperature of the whitewashed roof is lower than the outdoor air average value. However, whitewashing is hardly a permanent and viable proposal for heavy flat roofs. In wet areas it is easily washed out by heavy rains, and in dry areas it is quickly rendered ineffective by dust and dirt accumulation. A layer of light-coloured stone chippings may in the long run prove more rewarding. The application of an insulation layer is strongly recommended, while remembering the effect of its position on the thermal behaviour of the whole roof element.

(i) *An insulation layer above the structural concrete* This reduces the amount of heat flowing to the concrete slab during the day. Because of the high heat capacity of the mass of concrete, the reduced amount of heat that flows into it is absorbed there with a slight rise in temperature. Thermal insulation placed above the structural roof but below a dark-coloured waterproofing asphaltic membrane allows this upper layer to be overheated by preventing heat loss from its lower surface. This causes swelling and blistering of the asphalt, and possible evaporation of its volatile oils. If the insulating material is vapour-permeable, such as mineral wool or foamed concrete, the water vapour accumulates above it and beneath the waterproofing layer. As moisture condenses at night and evaporates during the day, pressure builds up and bubbles may form, tearing the waterproofing membrane loose from the slab below. For this reason a light external colour is essential in warm regions, even when the roof is well insulated and is designed with adequate provision for ventilation and removal of water vapour from between the insulating and waterproofing layers.

(ii) *An insulation layer below the structural concrete* Concrete absorbs great quantities of heat and, as its thermal resistance is low, the underside temperature of this layer closely follows the external surface pattern. Thus the upper surface of the insulation layer is at a much higher temperature than the indoor air. Despite the thermal resistance provided by the insulating material, enough heat flows through to raise the inner surface temperature, as the heat capacity of the insulation is very low. Therefore, the ceiling temperature and maximum heat flow to the interior are higher than when the insulating layer is external.

(*b*) *Pitched roofs*

Pitched roofs are usually of a simple truss type with a roof covering on its top chord, and with or without a suspended ceiling along its bottom chord. Again from research observations, it has been found that:

(i) in roofs composed of red cement tiles and ceiling of plastered expanded metal, ventilation of the roof space had no potential cooling capacity due to the effect of natural air movement through the tiles which did not fit tightly;

(ii) in roofs with corrugated galvanized steel covering and a ceiling of asbestos-cement sheets, attic ventilation reduced indoor temperature during the day by 0·5 to 1° only.

(*c*) *Single-layer lightweight roofs*

These are usually made of corrugated sheets of asbestos-cement, aluminium or galvanized steel, without additional ceiling materials, and the indoor climate is directly affected by fluctuations in the underside temperature of the roofing. Therefore, the thermal effect of these roofs during the day depends almost entirely on the external finish of the sheets, and ventilation of the interior is very important for the comfort of occupants.

When the external colour of the sheet is white, heating by solar radiation is almost prevented, while long-wave radiation to the sky is fully utilized. Under such conditions, the indoor temperature barely exceeds that of the outdoor air during the day, while at night it approaches the outdoor minimum. Provided that whitewashing can be frequently renewed, these conditions are satisfactory in some warm regions.

Ceiling height and human comfort

Ceiling heights in some parts of the tropics are exaggerated and their influence on the indoor climate tends to be over-estimated. A popular belief is that the higher the ceiling, the cooler the room. However, it is the ceiling temperature and not its height which is the decisive factor. Numerous studies have been carried out in various parts of the tropics, and the outcome is a consensus that the effect of height on the intensity of radiation is very limited and physiologically insignificant. Air temperature in rooms with ceiling heights of 2·5–3 m varied by less than 0·5°C.

In multi-storey building, reduced ceiling height in all except the upper storey results in reduced indoor air temperature. But in a two-storey building, low ceiling heights (2·5 m) have a slight thermal advantage over the higher ones (2·9 m). In top-storey apartments, the influence of ceiling height is dependent on type of roof. There is no adverse effect, either climatologically or psychologically in a reduction of ceiling height from 2·40 m to 2·25 m.

Indoor air motion is governed by the design of the openings only, irrespective of ceiling height.

However, ceiling height has an indirect influence on ventilation (stack effect) and on lighting, but this has been found quantitatively insignificant, as both of these depend mainly on other factors such as area, position and orientation of openings, and the reflectivity of surfaces to light rays, as well as on the microclimatic conditions.

Accordingly, ceiling height may be designed and selected on economic or purely psychological grounds. To relate this to the human scale, it is suggested that the adequate minimum height is that given by Le Corbusier of 2·26 m which can barely be reached from a standing position, but this varies among nations according to people's average height. Individuals differ in their notions of comfort and space, and many prefer the 2·70 to 3·0 metre range, even without a heat problem.

Urban planning and microclimate (see also chapter one for a more general appraisal)

Traditionally, man knew that every elevational difference, the character of land cover and every water surface, induces variations into the local climate, and he was careful in siting his settlements. However, these man-made settlements, or built environments, can create microclimates of their own, deviating from the macroclimate of the region in a way that depends on the extent of man's intervention. Such intervention with the natural environment is becoming greater as urbanization proceeds in the tropics with the growth of new towns and cities. Thus, consideration should be given to the urban microclimate, as climatic deviations play an important part in architectural land utilization and building solutions.

The deviations occurring in an urban climate (from that of the regional macroclimate) are usually caused by the following factors:

(*a*) *Changed surface qualities.* Plant and grassy covers reduce temperature by absorption and cool by evaporation; buildings and pavements increase solar radiation absorption and reduce evaporation.
(*b*) *Buildings* store the absorbed day heat in their mass and slowly release it at night; they also cast shadows and act as barriers to the wind. Wind channelling can cause localized increase in air velocity.
(*c*) *Energy seepage* occurs through the walls and ventilation of heated buildings; the heat output of industry, motors and electrical appliances may be appreciable.

(*d*) *Atmospheric pollution.* Waste emissions, fumes and vapours from domestic and industrial use will reduce direct solar radiation but increase the diffused radiation and provide a barrier to outgoing radiation; the presence of solid particles in the urban atmosphere may assist in the formation of fog and induce rainfall.

These deviating factors can affect the urban environment quite significantly. For instance:

Air temperature in a city may be 8–10°C higher than in the surrounding countryside.

Relative humidity is reduced by 5–10% due to the quick run-off of rain water from paved areas, the absence of vegetation and higher temperature.

Wind velocity can be reduced to less than half of that in the adjoining countryside, but the funnelling effect along a closely built-up street or through gaps between slab blocks can more than double the velocity.

Thus, urban planning affects several environmental factors which are important to the comfort and well-being of the inhabitants. Careful consideration should therefore be given to the pattern of sun and shade; the degree of protection from radiation, rain and wind; the ventilation conditions. These, in turn, will be influenced by: the dimensions and particularly the heights of buildings, the spacing of buildings, the variation in sizes and heights in any one section of a town, the orientation of the street network, and the distribution and extent of open spaces and gardens. In a warm humid climate, planning should be directed towards optimization of ventilation conditions and providing the maximum protection from solar radiation. On the other hand, in hot dry areas, the main consideration is to reduce the impact of solar radiation on buildings and to provide shade in the streets and open spaces. Where hot dry winds are associated with dust or dust storms, wind control should be aimed at protection rather than necessarily obtaining the best ventilation.

The successful building solution will be that in which the designer has fully accepted and understood the potential benefits that can accrue from harnessing and controlling nature's forces and using them to the advantage of human comfort. To achieve this he must be fully aware of the knowledge that has been developed in the fields of

(*a*) air flow around buildings,
(*b*) ventilation and air movement inside buildings,
(*c*) effects of landscaping and open spaces.

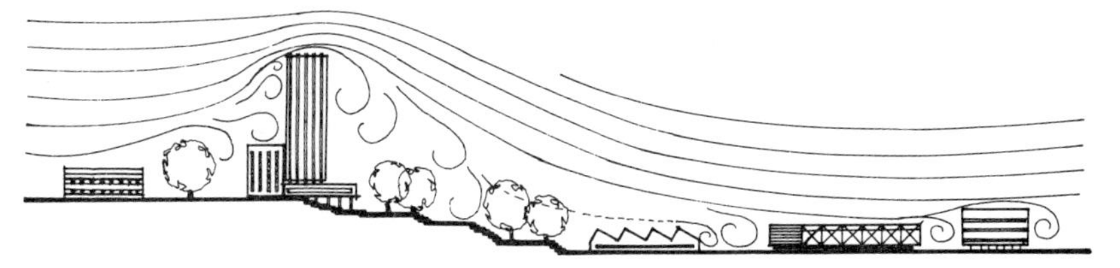

Figure 2.20 Air flow around buildings

Air flow around buildings

The main effect of a building block's dimensions is on the ventilation of nearby buildings and their exposure to sunshine and shade. A building in an open environment will create a wind shadow on its leeward side, where the wind velocity is less than that on the windward side.

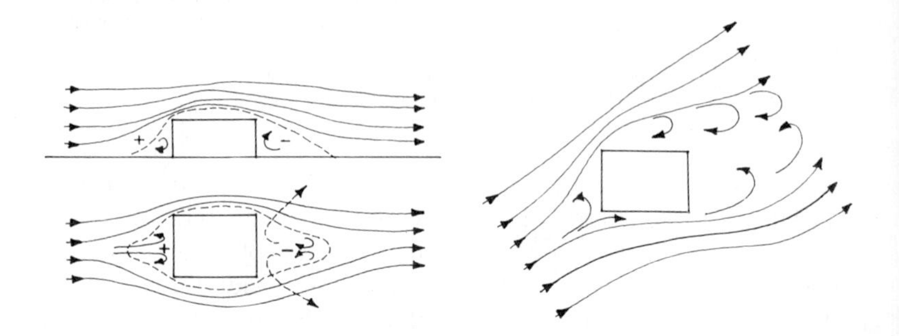

Figure 2.21 (*a*) Wind acting at right angles to face of building. (*b*) Wind acting at an angle to building face.

Thus, the rows of buildings facing the wind cause a reduction in air velocity around buildings behind them. This generally results in the wind velocity in a built-up area being much lower than in open country. Streets and open spaces enable the passage of wind among the buildings and help to improve ventilation conditions in the inner parts of a town. However, care should be exercised in the layout of streets as they affect the siting of buildings. A linear layout with a grid-iron building arrangement can produce wind shadows with stagnant air zones. A spacing of six or seven times the building height would be necessary to ensure adequate air movement for the following rows.

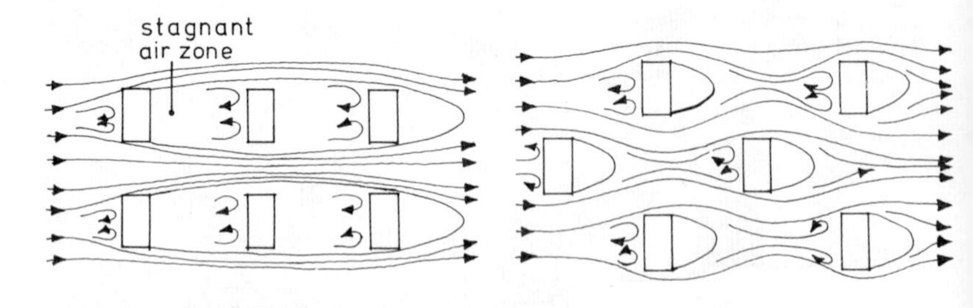

Figure 2.22 (*a*) Grid-iron layout (*b*) Checkerboard layout

The use of a checkerboard layout would allow a more uniform air flow and help almost to eliminate the stagnant air zones.

Air stream separation at the face of a building causes more air flow above it, which in turn induces a secondary flow at the lower built-up area. This effect increases along streets and in open spaces. Where the air stream separates on the face of a tall block, a vortex forms in front and others are shed to each side, causing an increased velocity at ground level and at the sides of the block (figure 2.23 (*a*)). This serves a useful purpose in hot climates.

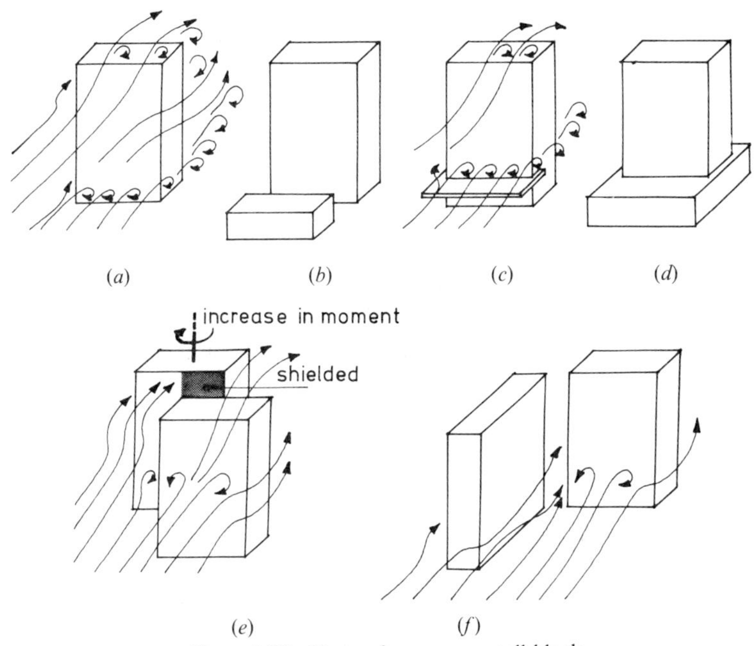

Figure 2.23 Vortex forms near a tall block

The vortex which forms on the windward side of the building and streams behind it could be strengthened by a low building positioned to windward (*b*). However, to reduce the strength of the vortex a canopy (*c*) or a plinth (*d*) could be used. Tall buildings are frequently constructed in groups, with the result that a downstream building may be shielded by an upstream one. This increases the turning moment on the leeward building, and may also increase local wind velocities due to reduction of pressure on the shielded area (*e*). If two tall buildings are positioned close together at right angles (*f*), wind velocities of up to 80% higher than the free stream can be expected in the gap between. This wind velocity is little affected by gap size or block width, but it is increased by any increase in the height of the buildings.

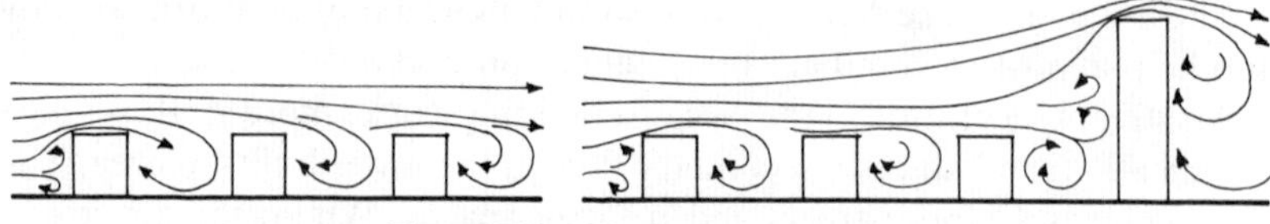

Figure 2.24 Air-stream separation

Single buildings projecting above the height of the neighbouring buildings modify appreciably the pattern and velocity of the air flow near the ground.

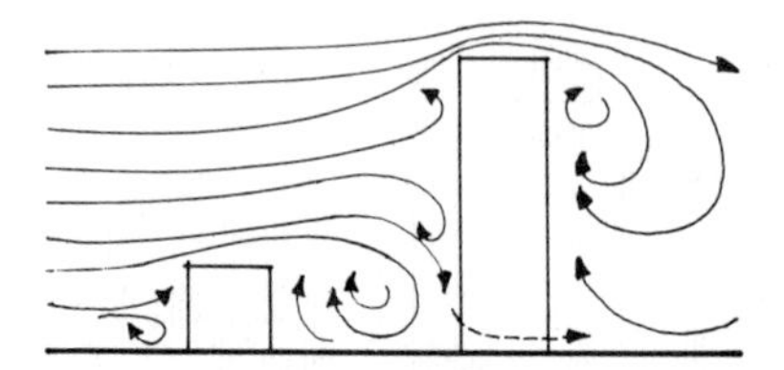

Figure 2.25 Air flow distribution about a high building with a low building to the windward side

When a low building is located in the wind shadow of a tall building (figure 2.26), the increase in height of the tall block increases the reverse air flow in a direction opposite to that of the wind through the low building. This is usually caused by the lower part of a large vortex returning through the building.

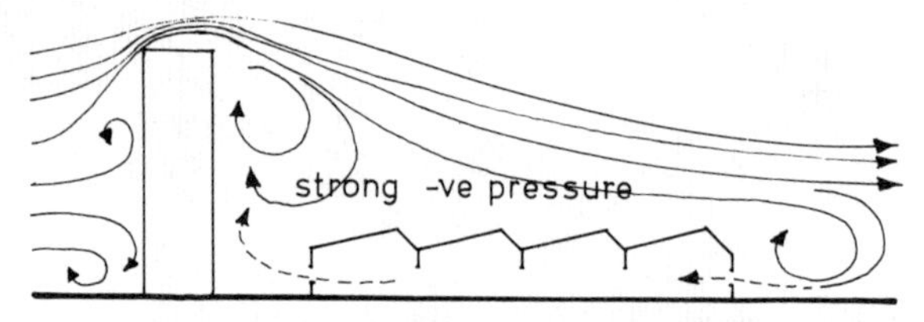

Figure 2.26 Reverse flow behind a tall block

In tropical climates air flow has particular importance as a remedy against high temperatures and humidities. Town layouts provide possible arrangements that may give wind protection and will utilize air movements beneficially through the proper building orientation.

Adaptation for wind orientation may not be of great importance in low buildings where the use of windbreaks, the arrangement of openings in the high and low-pressure zones, and the directional effect of window inlets can help to ameliorate the air flow. However, for high buildings, where the surroundings have little effect on the upper storeys, careful consideration has to be given to wind orientation.

Wind velocities are much higher above the level of the bulk of the building blocks in a town, so that higher buildings are exposed to higher winds. In hot climates, particularly humid ones, this may be an obvious advantage. In hot-dry regions having sand or dust storms, the higher floors will be exposed to the lesser concentration of sand, as this drops sharply in most cases above 10 m from ground level.

Ventilation and air movement inside buildings

A minimum level of ventilation is required to provide fresh air and remove carbon dioxide, odours and excessive water vapour. Its magnitude depends on the uses to which a room is being put, and the number and habits of its occupants. Particularly in warm-humid climates, air movement is required for comfort. This may be provided by fans or by natural means through ventilation. The functions of ventilation are therefore:

(*a*) *Supply of fresh air.* This is governed by the type of occupancy, number and activity of occupants, and by the nature of any processes carried out in the space. Requirements may be stipulated by building regulations and advisory codes in terms of m^3/h per person or air changes/hour, but these are only applicable to mechanical installations. However, even in a closed building, three air changes per hour can occur due to normal air leakage. A slight breeze could account for a ventilation rate of about 30 air changes per hour, while a stiff breeze may produce about 300 air changes per hour in a building.

(*b*) *Convective cooling.* The exchange of indoor air with fresh outdoor air can provide cooling if the latter is at a lower temperature than the indoor air. The moving air acts as a heat-carrying medium.

Ventilation (supply of fresh air and convective cooling) involves the movement of air at a relatively slow rate. The motive force can be either thermal or dynamic (wind).

(*c*) *Physiological cooling.* The movement of air past the skin surface accelerates heat dissipation in two ways:
(i) increasing convective heat loss,
(ii) accelerating evaporation.

Both the bioclimatic chart and the ET nomograms (figures 2.11 and 2.10) show the cooling effect of air movement, and how much higher temperatures can be tolerated with adequate air velocity. In very low humidities ($<30\%$) this cooling effect is not great, as there is an unrestricted evaporation, even with very light air movement. In high humidities ($>85\%$) the cooling effect is restricted by high vapour pressure preventing evaporation, but greater velocities (1·5–2m/s) will have some effect. In medium humidities (35–60%) it is most significant.

Indoor Air Flow Patterns When the width of a building is greater than the depth of its rooms, one room must be ventilated in conjunction with others, either by a direct connecting opening or through an intermediate space. If a flat comprises interconnecting rooms, the incoming air stream may have to change direction a number of times before leaving through the outlet, and these changes impose a higher resistance on the air flow. It is also important to provide openings on more than one side of a building to prevent wind buffeting.

In tropical climates cross-ventilation is extremely desirable, and openings should be provided on opposite sides of the building.

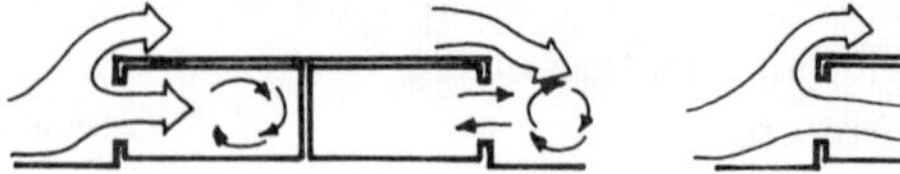

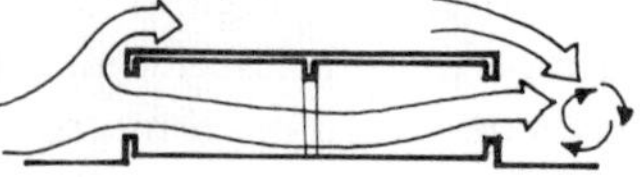

Figure 2.27 (*a*) Windward or leeward opening only; pressure build up on windward side and buffeting may occur in both spaces. (*b*) A connecting opening needed for cross-ventilation and to prevent buffeting.

Many experiments have been carried out to establish the effect of internal sub-divisions of a space on internal air velocities and on flow patterns. Givoni established that on the whole sub-divisions reduce the internal velocities moderately, the greatest reduction in average speed being from 44·5% to 30·5%. When the partition is in front and near to the inlet opening, the velocities are lowest, as the air has to change direction upon entering. Better conditions are obtained when the partition is located nearer the outlet opening.

We can therefore infer that satisfactory ventilation is possible in spaces where air has to pass from one room to another. It is also preferable for the upwind room to be the larger in size, to facilitate cross-ventilation through the building.

Other factors affecting the indoor air flow pattern and velocity should also be considered.

Wind Orientation Pressure on the windward side of the building is greater when the wind direction is perpendicular to the face of the opening. Wind at 45° produces a drop of about 50% in pressure, resulting in reduction of internal air velocity.

External Features The external features of a building can influence the wind flow pattern, which in turn affects the internal air velocity. For instance, if the wind is acting at 45° to an elevation, the projecting part of an L-shaped-plan building can more than double the positive pressure build-up. Any extension of the elevational area facing the wind will result in the increase of the pressure build-up. If a gap between two buildings is closed by a solid wall, a similar effect will occur.

Location of Opening The provision of openings on both the pressure and suction sides (windward and leeward) of a building produces internal air velocity almost three times that produced from openings of the same total area but located on one side only. From the thermal aspect, orientation of openings is of no importance as long as the openings are adequately shaded against the solar thermal load.

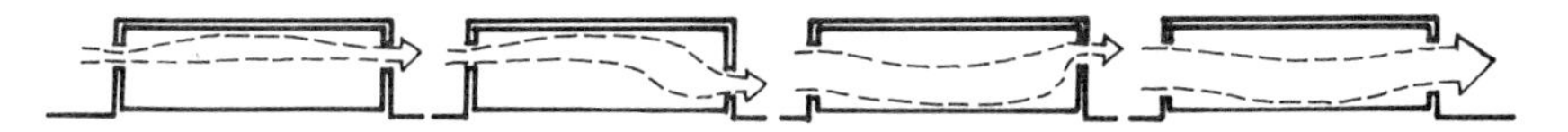

FIgure 2.28 Location of opening

Air movement must be directed at the body surface, i.e. it must be mostly through the living zone (1·20–2·0 m high) (figure 2.28).

The relative magnitude of pressure build-up in front of a solid area of the elevation (which in turn depends on the size and position of openings) governs the direction of the indoor air stream, and this is independent of the outlet opening position.

A larger solid surface creates a larger pressure build-up, and this pushes the air stream in an opposite direction, both in plan and section. As a result, in a two-storey building the air flow on the ground floor may be satisfactory, but on the upper floor it may be directed against the ceiling. One possible remedy is an increased roof parapet wall (figure 2.29).

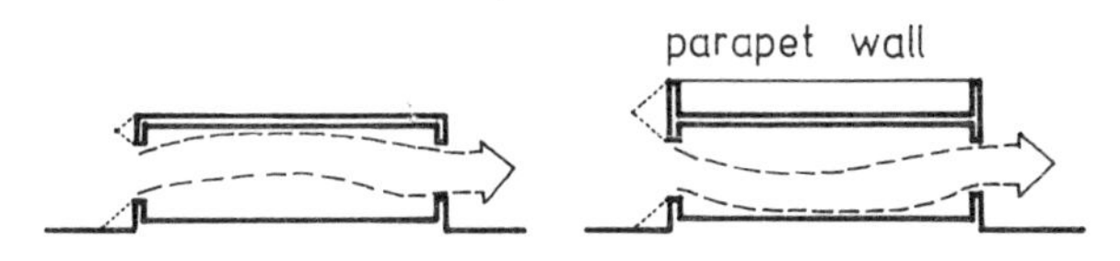

Figure 2.29 Direction of the indoor air stream

Size of Opening With a given elevation area, the largest air velocity is obtained through a small inlet with a large outlet. This is partly due to the total force acting on a small area, forcing air through the opening at a high pressure (wind force = pressure × area) and partly due to the Venturi effect in the imaginary funnel connecting the small inlet to the large outlet; the sideways expansion of the air jet further accelerates the particles.

Such an arrangement is useful when the air stream is to be directed at a given part of the room. When the inlet is large, the air velocity through it will be less, but the rate of air flow (volume of air passing in unit time) will be higher. When the wind direction is not constant, or when the air flow through the whole space is required, a large inlet opening will be preferable. The best arrangement is full wall openings on both sides, with adjustable sashes or closing devices which can assist in channelling the air flow in the required direction following the change of wind.

Controls of Openings Sashes, canopies, louvers and other elements used to control openings also influence the indoor air flow pattern and velocity. Sashes can divert the air flow upwards, while a reversible pivot sash channels it downwards into the living zone. Canopies above windows can eliminate the effect of pressure build-up and result in an upward air flow to ceiling. To ensure a downward air flow through the room, a gap is left between the canopy and the building face. Louvers and shading devices could also present a similar problem. The blade's position should not exceed the range of 15° downward slant to 20° upward slant if the air flow is to be directed to the living zone.

Protection against flies, mosquitoes and other insects is necessary in tropical areas. But the provision of fly-screens can cause considerable reduction in air flow through openings, especially if the external wind is slow. The decrease in total air flow caused by a 16-mesh 30-gauge wire screen is about 55% for a wind of 0·75 m/s, and only about 25% when the speed is 4·5 m/s. For nylon netting, the reductions are 35% and 46% for wind speeds of 0·75 and 3·80 m/s.

The reduction in internal air speed due to screens over a single central window is greater with an oblique wind than with a perpendicular wind. On the other hand, this difference is not observed with lateral inlets.

Application of screens to a whole balcony in front of the openings improves on the ventilation conditions obtained with screens directly applied to the windows. The wind is able to penetrate the fly-screen through a large area, and then to contract towards the smaller window, free from obstruction. Application of screens to the outlet window or rear balcony produces a smaller effect than the front screens. This supports the hypothesis that the fly-screen does not merely give additional resistance to air flow, but causes the wind to slip over it, preventing the initial entry of the air, besides reducing the speed of that air which does pass through.

Exclusion of rain is not difficult, and making provision for air movement does not create any particular difficulties, but together and simultaneously these can create problems. Opening of windows during periods of wind-driven rain admits rain and spray, while closing the windows creates intolerable conditions indoors. The conventional tilted louver blades are unsatisfactory for two reasons:

(i) Strong wind drives the rain in, even upwards through the louvers.
(ii) The air movement is directed upwards from the living zone.

Verandahs and large roof overhangs are the best traditional methods of protection.

The Stack Effect The stack effect relies on thermal forces set up by density difference (caused by temperature differences) between the indoor and outdoor air. It can occur through an open window when the air is still, the warmer and lighter indoor air flowing out at the top, and the cooler denser outdoor air flowing in at the bottom. Special provision can be made in the form of ventilating shafts and air scoops. The higher the shaft, the larger the cross-sectional area and the greater the temperature difference; the greater the motive force, the more air will be moved.

The air scoop or "Malkaf" used in most rural areas of Egypt fulfils various functions, from controlling air supply to filtering out sand and dust, and the provision of evaporative cooling and humidification. This is necessary, as in hot dry climates the building has to be closed during the day to preserve within the high-thermal-capacity structure the cool night air retained from the previous night.

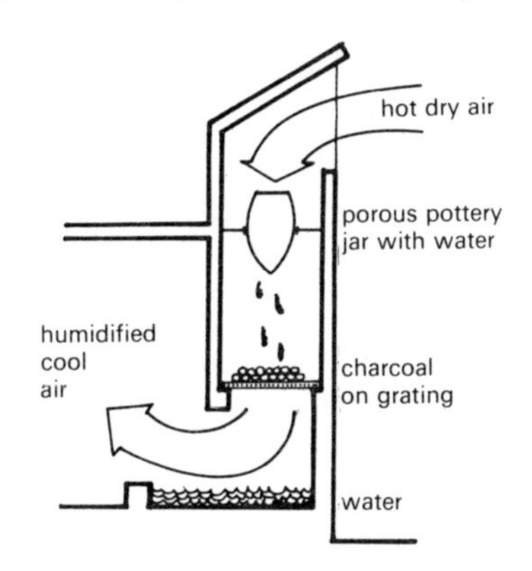

Figure 2.30 "Malkaf" or air scoop

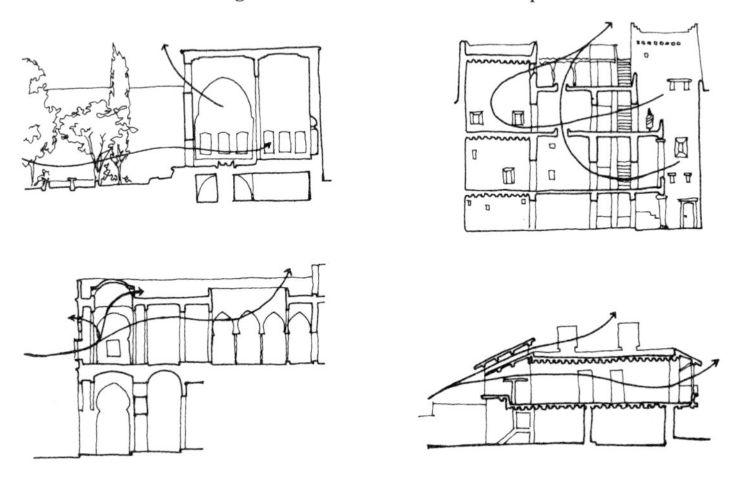

Figure 2.31 Natural ventilation in Islamic architecture (after Abdulak)

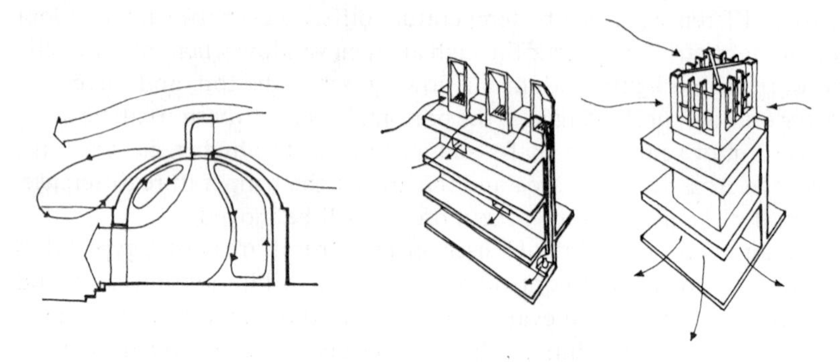

Figure 2.32 Air scoops (after Abdulak)

In some parts of India, raffia curtains are hung in front of the openings on the windward side. They are moistened with water from time to time, allowing the air passing through this moist loosely textured curtain to be cooled and humidified.

Effect of landscaping

The immediate surroundings of low structures have definite effects on air-flow patterns and on wind velocities. The landscape design elements, including plant materials, trees and shrubs, walls and fences, can create high and low pressure areas around a building. Care should be taken that arrangements do not eliminate the desirable cooling breezes during hot periods, and planting should be designed to direct and accelerate beneficial air movements into the buildings. Besides their aesthetic and shade-giving properties, the value of tree windbreaks lies in their ability to reduce wind velocities. This mechanical effect brings perceptible changes, both in the temperature and humidity of the air in evaporative effects.

A windbreak diverts the air currents upwards and, while they soon turn back and again sweep the ground, an area of relative calm is created near the ground. The type of windbreak used has a definite effect on the resultant air-flow pattern and on the area of protection. Solid wind barriers (most commonly used) or walls cause eddies over the top which reduce their effectiveness, while tree belts with greater density and thickness produce a larger effect in wind protection (Table 2.5 and figure 2.33).

Table 2.5 Effect of windbreaks

Object	*Height*	*75% reduction*	*50% reduction*	*25% reduction*
wall	H	13·0H	15·5H	21·5H
trees	H	—	13·5H	27·0H

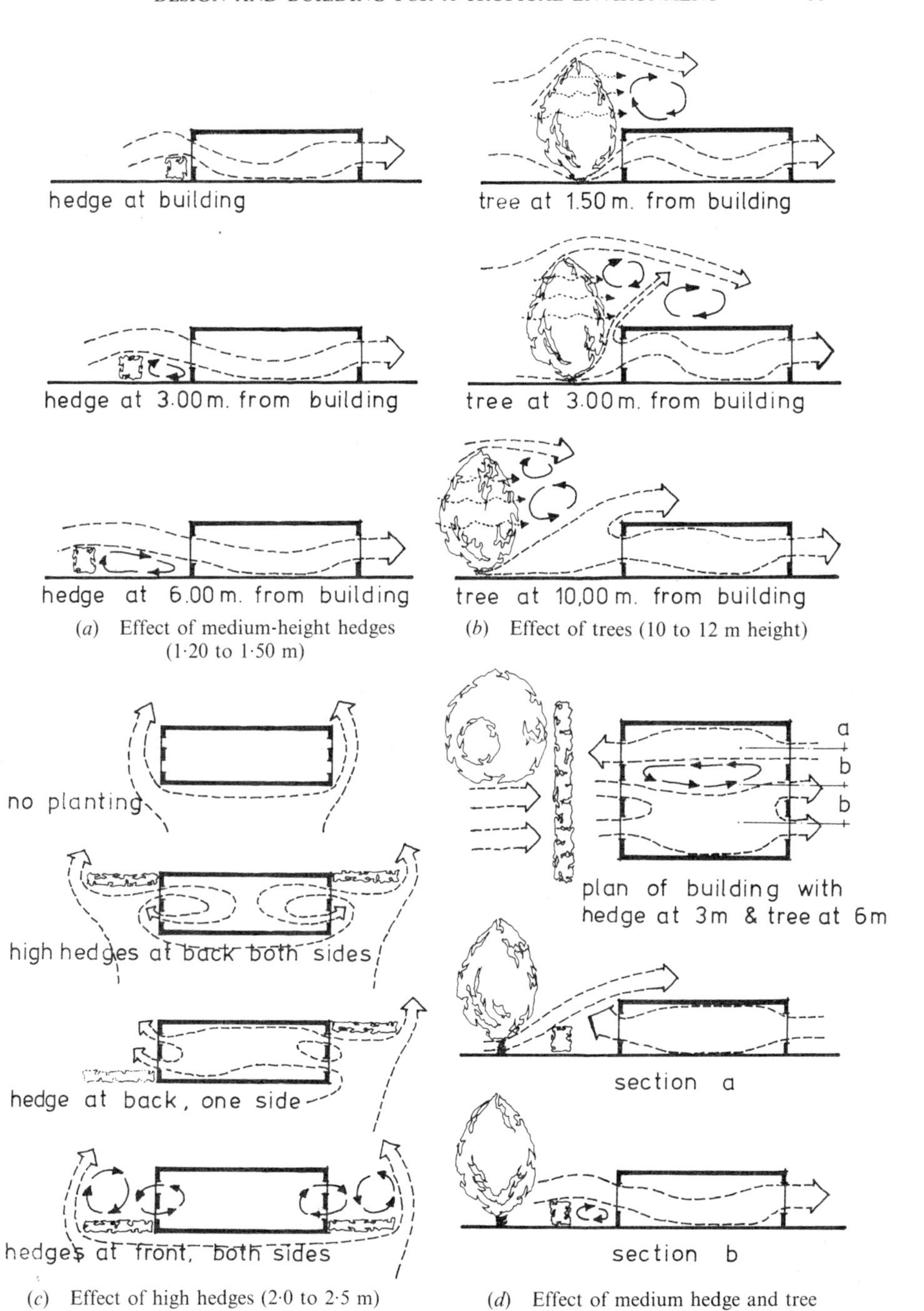

(*a*) Effect of medium-height hedges (1·20 to 1·50 m)

(*b*) Effect of trees (10 to 12 m height)

(*c*) Effect of high hedges (2·0 to 2·5 m)

(*d*) Effect of medium hedge and tree

Figure 2.33 Modification of air flow pattern with landscaping

Although there is no 75% reduction for trees because of the jet movement of air through them, they cause a more extended area of protection.

Building design in tropical climates

The building design considerations relating to climatic control are considerably different in cold and in hot conditions. The thermal effect of materials in buildings without mechanical air-conditioning in hot climates is primarily dependent on the diurnal temperature range (which in turn depends mainly on the level of vapour pressure). Thus the air temperature and the humidity combine in determining the type of climate. Prevention of water penetration in buildings is a requirement in all rainy regions.

Using the tropical classification of climates proposed by Atkinson, the following is a thermal design approach appropriate to each zone. This approach is developed from each of the different climatic characteristics, through the human requirements, to principles of design and details for building construction.

Warm-humid equatorial climate

This climate zone is characterized by high ambient temperatures (21–32°C), high humidity, high and fairly evenly distributed rainfall, small diurnal and annual variations of temperature, little seasonal variation, light winds, and long periods of still air. The reflected radiation from the ground is usually low, as the vegetation is dense and the damp soil is dark.

The physiological thermal requirements and hence the building characteristics are the same for the whole year, as the seasonal climatic variations are slight. The main cause of discomfort is the subjective feeling of skin wetness. Continuous ventilation is therefore required to ensure a sweat evaporation rate sufficient to maintain thermal equilibrium and minimum sweat accumulation of the skin. Radiant solar heat gain should be prevented.

Planning and layout should aim for

(*a*) elongated plan shapes with a single row of rooms to allow cross-ventilation
(*b*) open layouts and orientation of buildings with long axes in the east-west direction for maximum benefits from the prevailing breezes and the minimum effect from the sun
(*c*) locating buildings on high ground and on the sides of valleys
(*d*) the use of raised buildings on stilts and high-rise buildings
(*e*) the inducement of air movement and the avoidance of totally enclosed courtyards.

Landscaping of internal spaces should provide shade and allow for free passage of air. Planting and maintaining of grassed areas and trees minimize heat reflection and glare, and have evaporative cooling effect. Pergolas and light framing covered with climbing plants are also very effective. Open ground under buildings provides useful shaded outdoor areas.

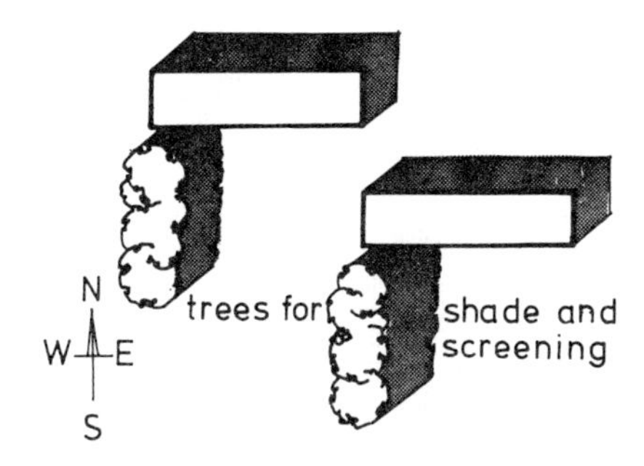

Figure 2.34 Open layout; long-narrow blocks with main rooms facing prevailing breeze

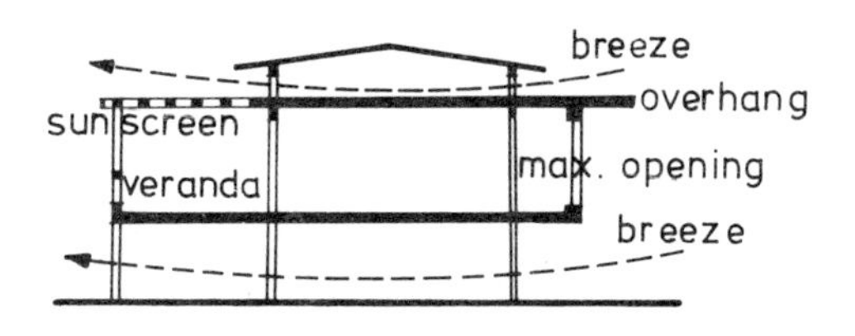

Figure 2.35 Raised building allows breeze to pass under for cooling

Roofs and walls should be constructed of low-thermal-capacity materials with reflective outside surfaces where not shaded. The roof must be of double construction, provided with a reflective upper surface and also a ceiling of highly reflective upper surface. The use of a good thermal insulation layer is recommended.

Openings should be as large as possible, and fully openable and placed to permit natural air flow through the internal spaces at living-zone level. Attention should be given to protection from driving rain and insects without drastically reducing air movement.

Examples of housing solutions are shown in figures 2.36 and 2.37.

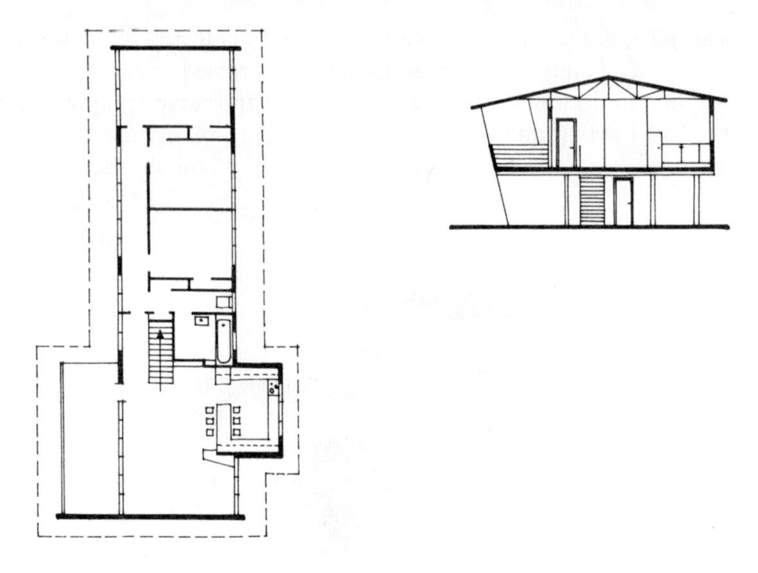

Figure 2.36 Housing type for Commonwealth employees, Darwin, Northern Australia. (after Koenigsberger)

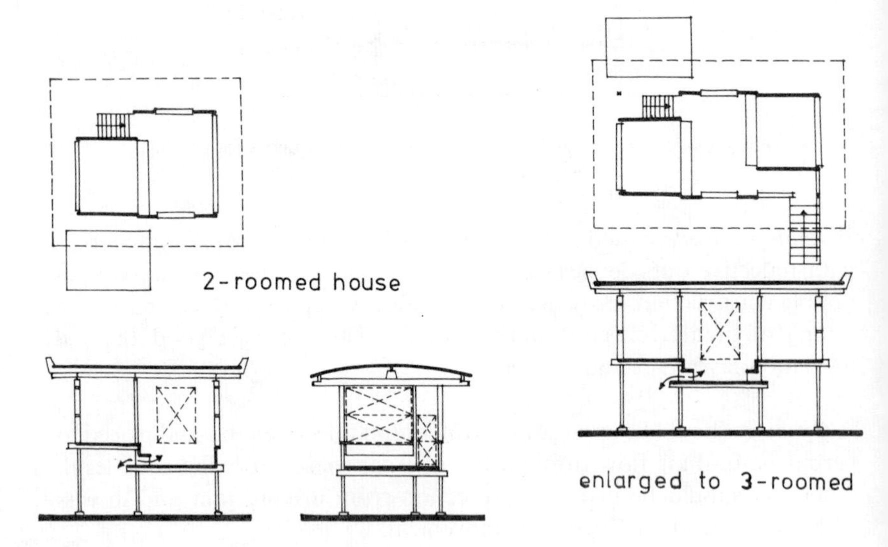

Figure 2.37 Experimental expandable type housing, Cambodia, utilizing traditional and system building methods.

Warm-humid island climate

This sub-group is a variety of the warm-humid equatorial climate and is slightly more favourable, with slightly lower temperature and a dominance of steady trade winds (6–7 m/s) of almost constant direction. This facilitates heat loss by convection and evaporation. However, as most of these islands lie in the tropical cyclonic belt, the structural and constructional details must be designed to withstand wind velocities of up to 70 m/s (force 12).

Hot-dry desert or semi-desert climate

This climate zone is characterized by high day temperatures and low night temperatures, low humidity and low precipitation, large diurnal temperature ranges, distinct seasonal variations between hot summers and cool winters, and little air movement except for local thermal winds, and sand or dust storms. Clear skies allow high intensity of direct solar radiation, augumented by reflected radiation from barren and light-colour terrain.

The physiological thermal requirements for comfort depend on reduction of intense solar radiation by day, and the reflected radiation from the ground and surrounding buildings. Buildings adapted to summer conditions will generally satisfy winter requirements. Walls and roof construction must maintain the inner surface temperature at a level less than skin temperature by day, as breezes can only be used to advantage by night and if dust is filtered out. Natural ventilation during the day is unnecessary, as the prevalent low humidity allows adequate sweat evaporation with little air motion.

Planning and layout should aim for

(*a*) the protection of buildings and external living spaces from solar radiation and hot dusty winds
(*b*) an enclosed compact and inward-looking plan with separate rooms for day and night use
(*c*) larger dimensions of buildings to face north and south, as west is the worst orientation
(*d*) secondary non-habitable rooms could be used on east and west sides of a building
(*e*) narrow roads and streets, arcades, colonnades and enclosed courtyards in order to get maximum shading effect and coolness.

Landscaping of external spaces should provide for shading of verandahs and courtyards. The use of trees, loggias, pools and water fountains within courts and open spaces cools the air by evaporation, keeps dust down, provides shade, and gives visual and psychological relief.

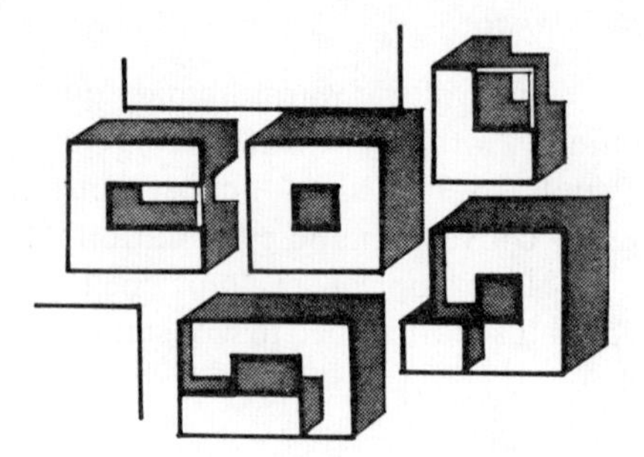

Figure 2.38 Compact layout and narrow streets; building around closed or semi-closed internal courts.

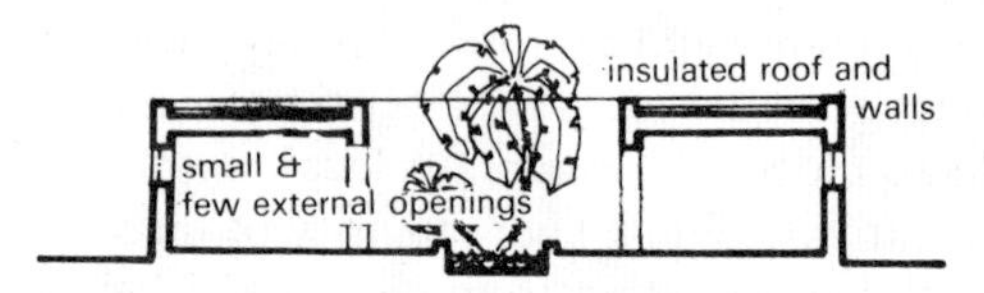

Figure 2.39 Main rooms face inward onto courtyard with planting and water ponds or fountains.

Roof and walls should be constructed of high-thermal-capacity materials to utilize the large diurnal temperature variations. Roofs must be constructed of heavy material with outside applied insulation and should slope towards courtyard. External surfaces of light colour will reflect a large part of the incident solar radiation and reduce heat entering the building fabric; dark-coloured surfaces should be avoided.

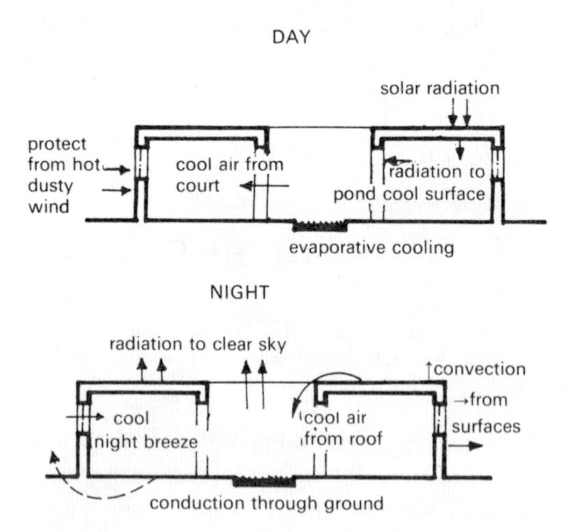

Figure 2.40 Thermal system of a small courtyard house

Openings should be small towards the outside, and large toward the courtyard. If large openings are used on external elevations, they must be provided with heavy shutters (usually wooden) of high thermal resistance. Internal or semi-internal courtyards with access to rooms through large openings must be protected, as for windows with moveable insulated shutters with a small aperture for illumination. During daytime, openings must be closed and shaded to keep ventilation to the minimum necessary for hygenic reasons. This is important to minimize entry of hot dusty air. However, at night ample ventilation is needed through opened windows to allow the cool night breeze to dissipate the heat stored in the structure during the day.

Examples of housing solutions are shown in figures 2.41–2.45.

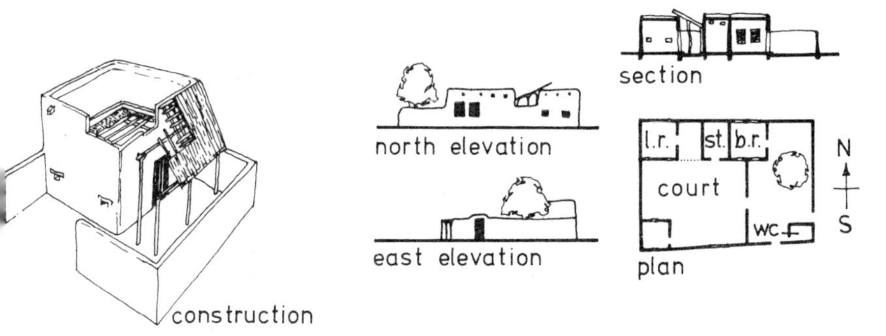

Figure 2.41 Jalouse and mud-brick peasant house, Sudan (after Danby)

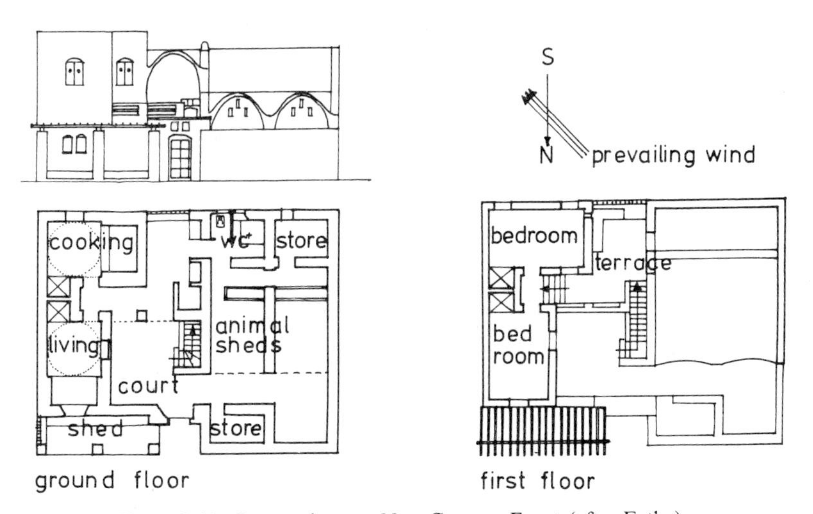

Figure 2.41 Peasant house—New Gourna, Egypt (after Fathy)

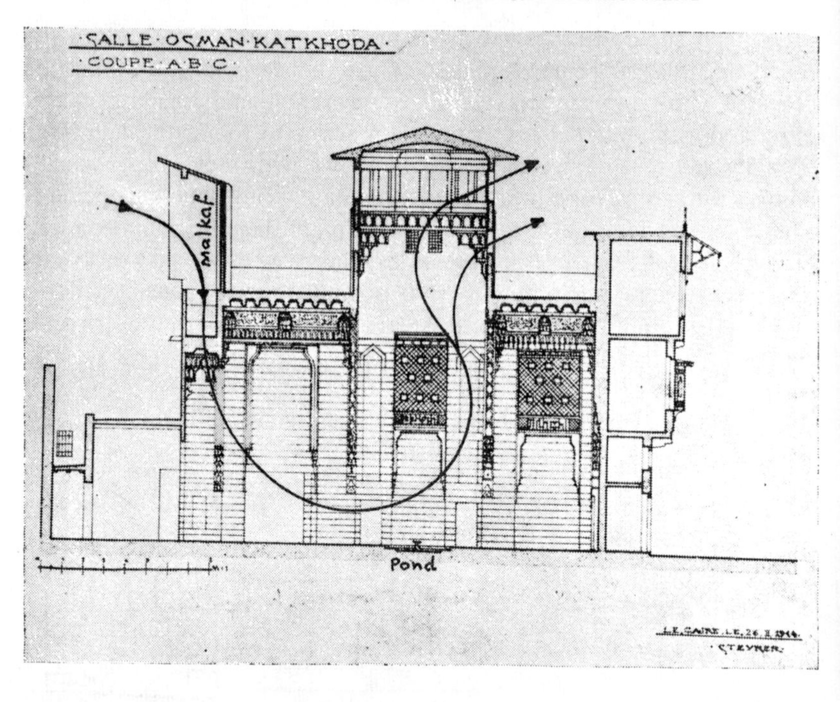

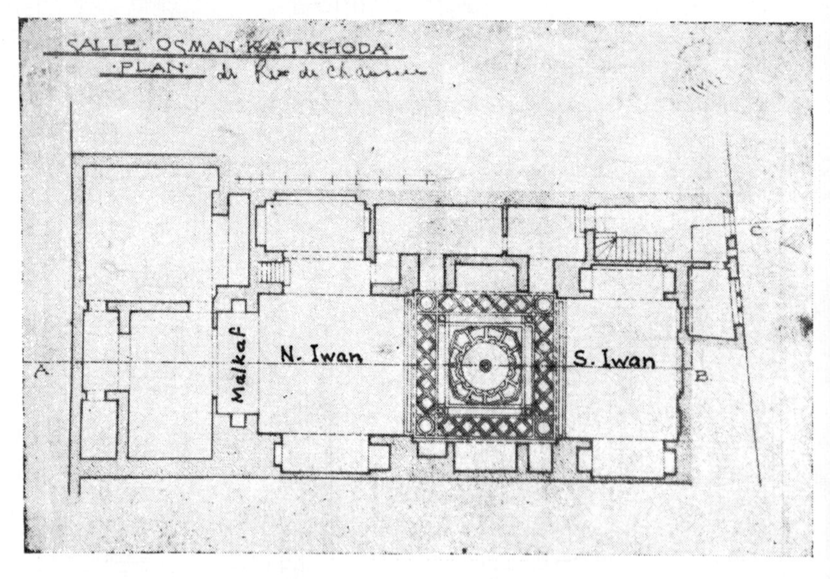

Figure 2.43 Plan and section of Osman Katakhda hall—Old Cairo, Egypt (after Sameh)

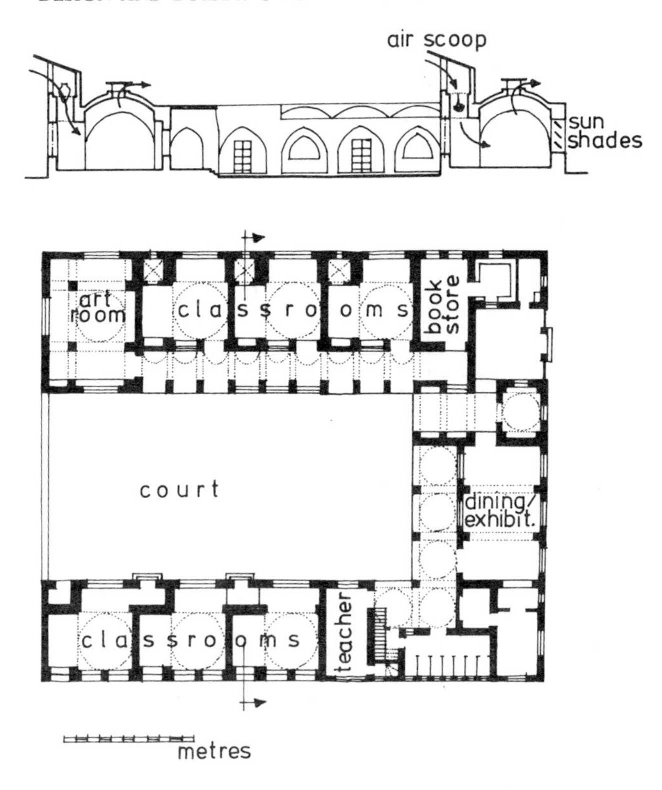

Figure 2.44 Plan and section of Primary School, New Gourna village—Egypt (after Fathy)

Hot-dry maritime desert climate

This sub-group is a variety of the hot-dry desert or semi-desert climate, and occurs where sea and desert meet. It is characterized by higher moisture content in the air, while percipitation is still low. This humidity tends to reduce the diurnal variation. The differential heating and cooling of land mass and sea produces alternating land and sea breezes. The night-time wind blowing towards the sea brings the hot inland desert air, possibly with sand and dust, and protection from it must be provided.

The buildings are usually of high-thermal-capacity walls and roofs, with no windows or very small ones, facing the inland direction. The side of the building facing the sea should be of lightweight construction with large openings. The various climatic responses of a courtyard house in Salala, Oman are shown in figure 2.46.

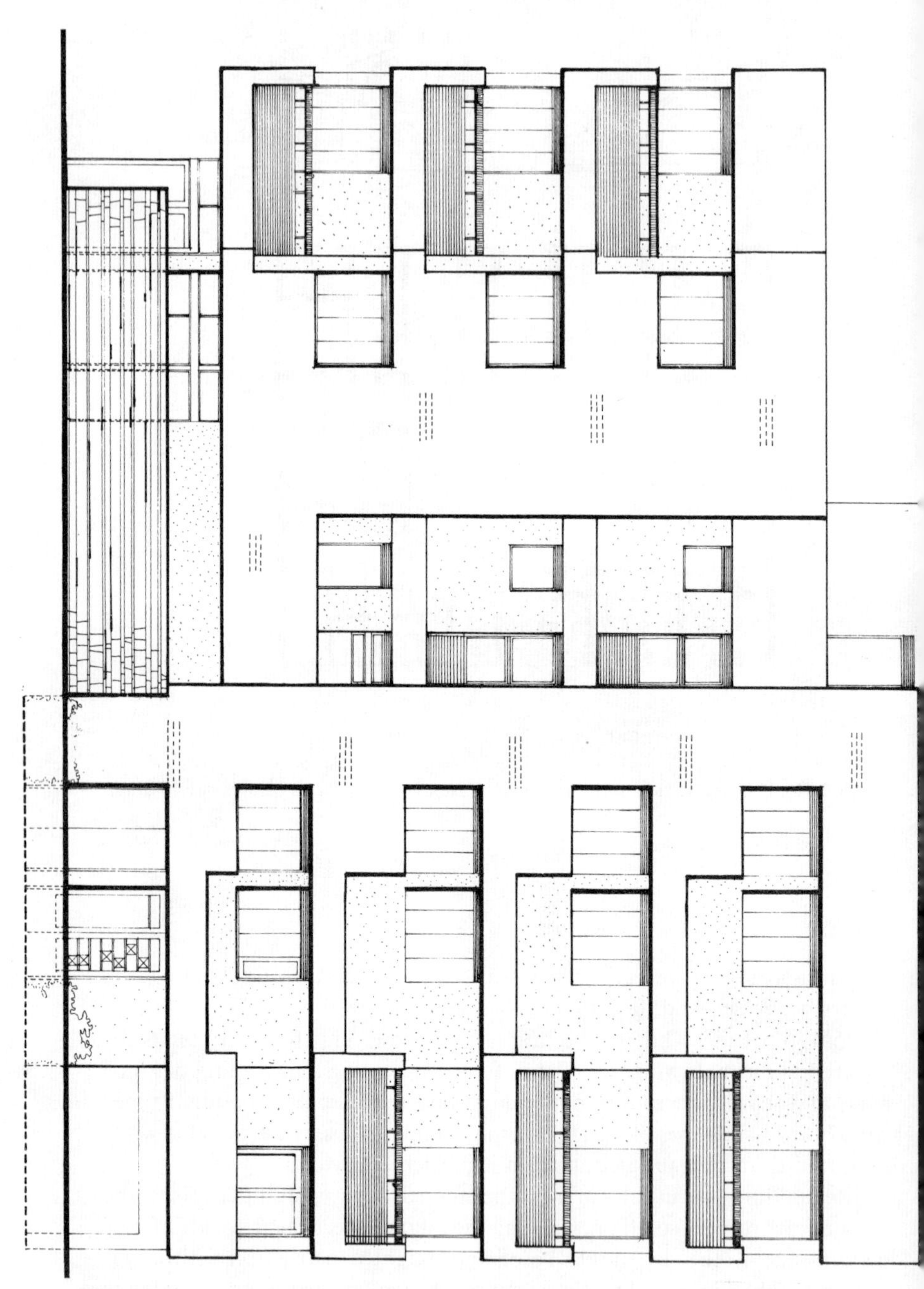

Figure 2.45 A typical plan (opposite) and elevation (above) of a block of flats—Cairo, showing the internal air movement for cross-ventilation.

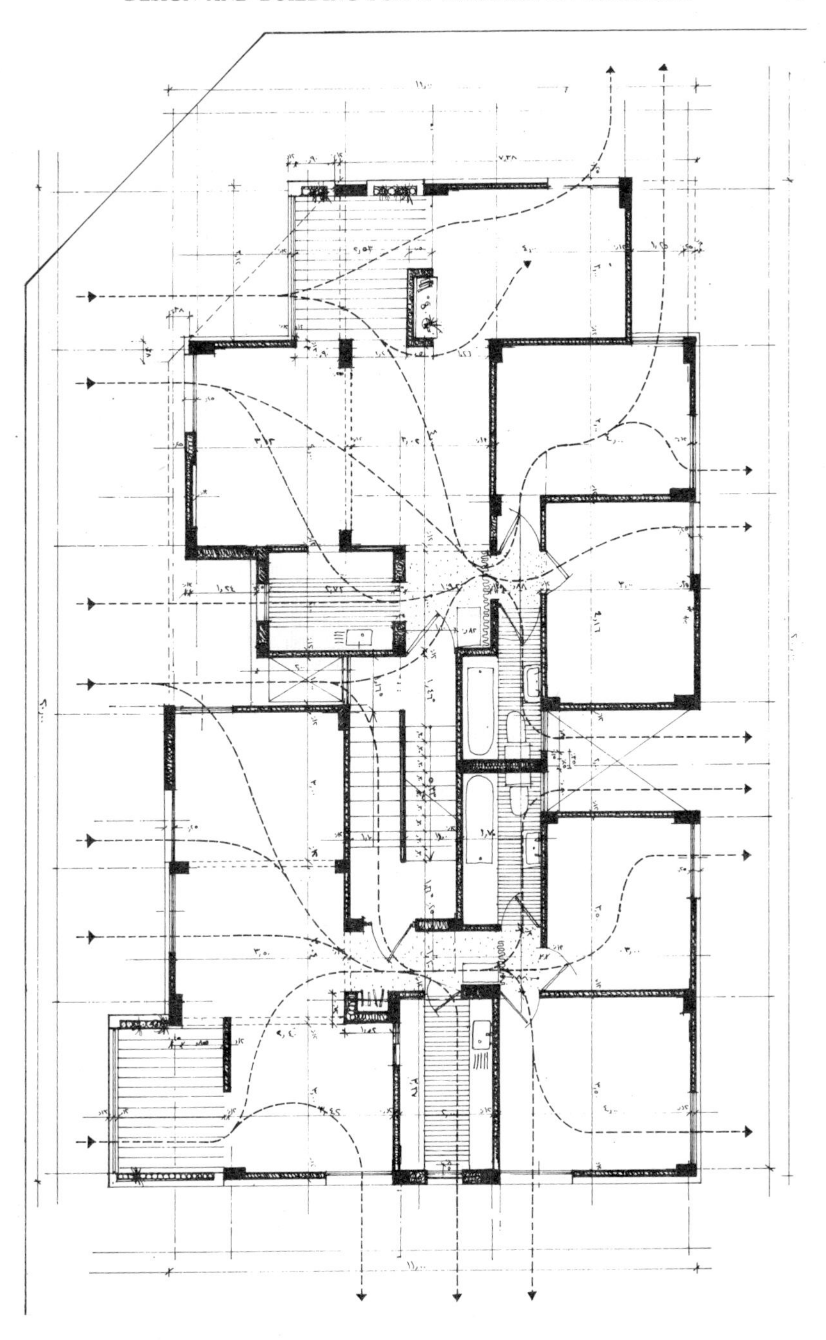

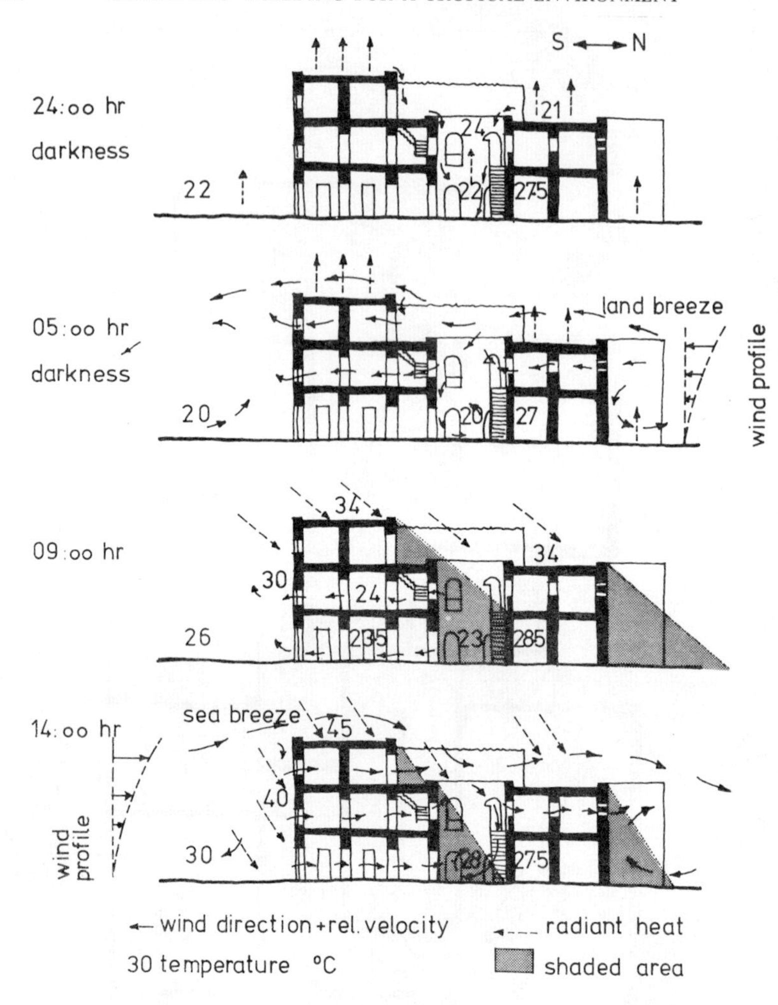

Figure 2.46 Climatic response of Salala courtyard house. Oct. 28, '73 (after Cain, Afshar, Norton)

Composite or monsoon climate

This climate zone is characterized by climatic changes from one season to another. There could be two, three or four seasons. One season may be similar to hot-dry desert, another to a warm-humid region, while a third may have cold nights and pleasant sunny and warm days with low humidity and little percipitation. Transitional periods of varying length may occur between the clearly discernible seasons.

The physiological thermal requirements are accordingly those of the warm-humid and the hot-dry climates as applicable to the various seasons of a composite climate. However, in the third cold season, human comfort will depend on preventing heat loss from the body, especially at night.

Planning and layout are as for hot-dry climates, for both hot and cold seasons but, for the monsoon or rainy season, buildings should be designed as for warm-humid climates. Where conflicting or incompatible requirements arise, priority must be given in accordance with the length of season and the severity of design conditions. Generally, a moderately compact internal planning of houses of the courtyard type is beneficial for most of the year. Houses with separate day and night rooms (as for hot-dry climates) are possible if the period is long. Buildings should be grouped to take advantage of prevailing breezes during the period when air movement is necessary. A moderately-dense low-rise development which ensures protection of outdoor spaces, mutual shading of external walls, shelter from wind in the cold season and from dust in the dry-hot season, is sensible.

Landscaping of external spaces must be controlled to provide protection against dust and thermal winds. Courtyards should be designed to allow sun penetration during winter months, and provide shade in the hot season. They may be covered by pergolas carrying deciduous creepers. Large projecting eaves and wide verandahs are needed in the warm-humid season as outdoor living areas, to reduce sky glare, provide shade, and keep the rain out. Louvers and sun shades used for protecting openings during the hot-dry period are also useful in the rainy season against rain and wind-driven spray.

Roofs and internal walls may be of high thermal capacity, allowing outer walls to have large openings. External walls and roofs should have an insulation layer placed on the outside. Surfaces exposed to the sun should be lightly coloured during the the hot season, and changed to a dark colour before the arrival of the cold season. This is valid in a large number of monsoon climate areas due to the economic labour cost of white or colour washes.

Openings should not be more than about 50% of the area of a wall when openings are on opposite sides, and about 25% if on adjacent walls. They should have solid shutters which can be opened for cross-ventilation during the warm-humid season and for night cooling during the hot-dry season.

Examples of housing solutions are shown in figures 2.47–2.49.

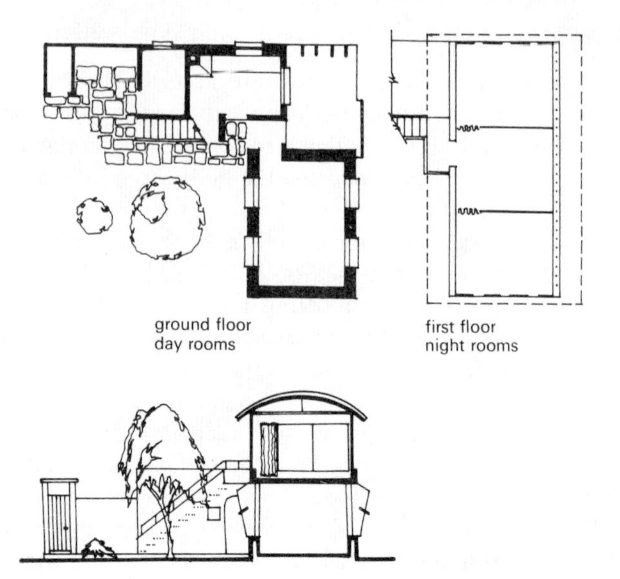

Figure 2.47 A house with separate day and night areas. (after Koenigsberger)

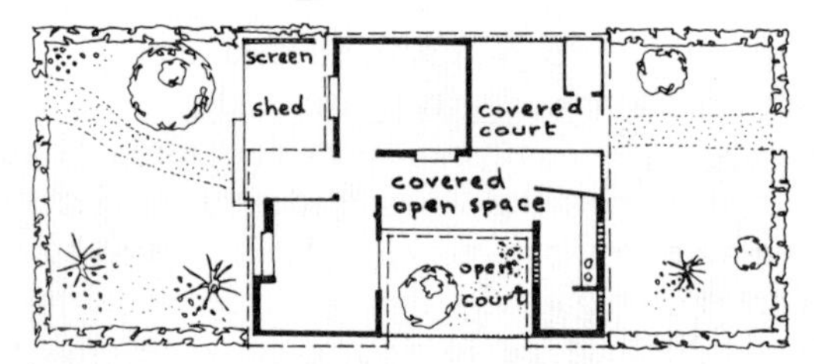

Figure 2.48 A house with a variety of external spaces.

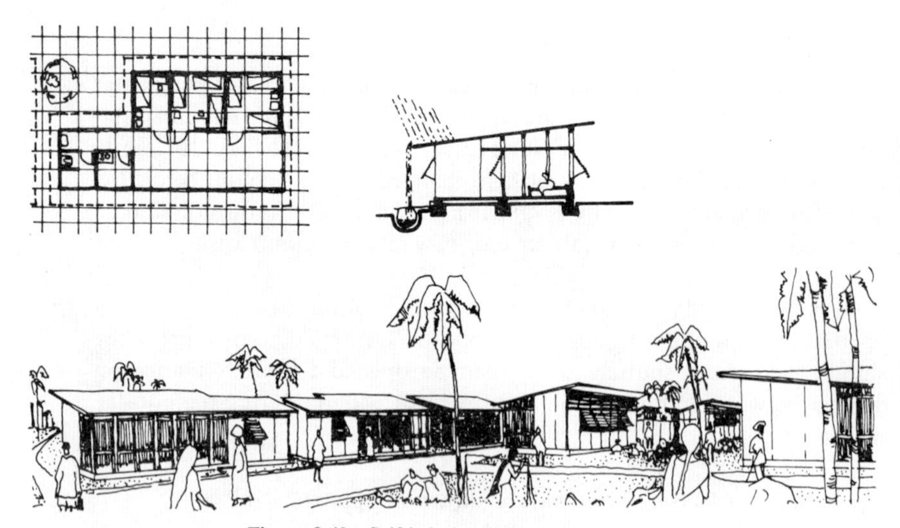

Figure 2.49 Self-help housing New Bombay

Tropical upland climate

This sub-group is a variety of the composite climate, but is further characterized by its distinct rainy seasons and its domination by solar radiation, often with moderate to cool air temperatures. Cold discomfort may occur at night, even in the warm season, and may require some mild heating to achieve comfort.

Buildings of oblong plan should be compact, with north-south elevational orientation. This allows minimum openings on the east, and particularly the west, where the solar heat gain is largest. Openings must be protected from the intense solar radiation, which often makes these solar control devices a prominent feature of the design.

As many living activities take place out-of-doors, well shaded external spaces are essential. The shading devices may be adjustable awnings or removable to allow sunshine in during the cooler period of the year.

Generally, buildings with the following criteria are suitable: walls of high thermal capacity and a minimum of 8-hours time-lag; in multi-storey buildings, well-insulated lightweight construction may be used, provided all openings are well shaded; window orientation is governed solely by the aspect of solar heat gain; window size of about 20% of the elevational area of a room is adequate.

Conclusion

Architects are trained to design buildings to fulfil certain functions, giving full consideration to all design aspects, including site and environmental conditions. Building in tropical climates brings the additional problem of adequate cooling, which can certainly be dealt with by mechanical means, but in siting and designing tropical buildings the architect should give serious consideration to the well-tried traditional methods which have enabled occupants in the past to be comfortable with maximum economy. These traditional solutions should be studied but not copied, as they have been developed under conditions that have changed considerably through time.

Nor is it practicable to plan a building exclusively on economic, functional or formal grounds, and expect a few minor adjustments to give a good indoor climate. Unless the design is fundamentally correct in all aspects, no specialist can make it function satisfactorily. Climate must be taken into account when deciding on the overall concept of a project, on the layout and orientation of buildings, on the shape and character of structures, on the spaces to be enclosed, and on the spaces between the buildings.

The designer's job is multi-problem-solving. He must cope simultaneously with topographical, climatic, psychological, physiological, economic, social, functional, operational and structural problems, over and above the aesthetic and various other architectural design criteria. All that, without losing sight of questions of communication and the framework in which the architectural solution to the problem takes place, namely the general city and regional planning context of his work.

It is impossible to establish an order of priority in dealing with these problems. During the early stages of a project, an idea or design concept must be produced which does not preclude the solution of any of the problems, and which promises to solve a good number of them in a convincing and elegant manner.

In the tropics, it is not enough to know the basic principles which will assist in providing favourable conditions for life through natural means, and then to manipulate proper controls on the climatic characteristics affecting human comfort; it is also important to know and consider the cultural and traditional aspects of life in the particular region in which the building will be sited. For instance, the type of clothing is not only an important factor in achieving a degree of comfort under difficult environmental conditions; it is also an indicator of social, economic, cultural and religious aspects of a country's tradition and life.

Architecture is about people. It should be the totality of human experience, desires, hopes, knowledge and aspirations, translated through man's technology and environment to fulfil his functional needs and emotional requirements. When this is not so, it is not architecture: it is a fad or fashion and it will die. When architecture is responsive, it will survive through the ages.

Acknowledgement

The author wishes to acknowledge the help of Mr. H. Serageldin, B.Arch., in the preparation of the diagrams.

FURTHER READING

Abdulak, S. and Pinon, P. (1973), "Maison en pays Islamique—Modèles d'architecture climatique", *L'Architecture D'Aujourd'hui*, Mai/Juin.

Atkinson, G. A. (1953), "An introduction to tropical building design", *Architectural Design*, Oct.

Ballantyne, E. R. and Spencer, J. W. (1972), "Climate and comfort in a humid tropical area", *Build International*, Vol.5. No. 4.

Building Research Station (1974), "Building for comfort—the architect's approach to design for comfort under tropical conditions", *Overseas Building Notes*, No. 158, Oct.

Cain, A., Afshar, F., Norton, J., (1974) "Indigenous building and rapid urbanisation—a case study of Salala in Southern Oman", *Architectural Assosiation Quarterly*, Vol. 6, Nos. 3–4.
Danby, M. (1973), "The design of buildings in hot-dry climates and the internal environment", *Build International*, Vol. 6, No. 1.
Givoni, B. (1969), *Man, Climate and Architecture*, Elsevier.
Huntington, E. (1948), *Civilisation and Climate*, Yale University Press.
Koenigsberger, Ingersoll, Mayhew, Szokolay (1973), *Manual of Tropical Housing and Building*: "Part I, Climatic design", Longman.
Oakley, D. (1961), *Tropical Houses*, Batsford.
Olgyay, V. (1963), *Design with Climate*, Princeton University Press.
Plant, C. G. H. (1972), "Windows: design and function under tropical conditions", *B.R.S. Overseas Building Notes*, No. 143, Garston.

CHAPTER THREE

TRANSPORT

ROBERT WHITE

Introduction

Origin of transport movements

All movement of people and goods, whether on foot or by vehicle, fulfils social, economic or military purposes. A household or community can exist without transport only if it is self-sufficient, and the nature and amount of resources to maintain that self-sufficiency will determine the standard of living. If a "primitive" society is defined as one lacking tools and technology, then such a society cannot have specialization of productive functions. When specialization and mass production arise then exchange of surpluses follows and transport movements are created. These statements are so obvious that the reader may question the need to make them but, outside professional circles, the problem has been discussed as though it were one of transport movements only. This is particularly true of controversy on the subject in recent years, when there has been more public concern with the nature of the urban environment. Discussion has ranged around whether the use of the private car in city centres should be restricted; whether road improvements are justified or necessary; or what new forms of mass transport might serve the movement of people in cities. Some years ago, discussion centred around an even narrower area, namely, the relief of congestion or the reduction in numbers of accidents by traffic measures and road improvements.

All discussion of the traffic problem in cities is unreal unless full account is taken of the effects of changes in the transport system on the way in which people live and work. Equally, full account has to be taken of the effects of changes in the way people live and work on the transport system.

Problems and conflicts

It is not unnatural that public concern with the effect of transport on the built environment should be directed primarily to cities and large towns, and should be most concerned with the effect of the motor vehicle. In Britain nearly 60% of the mileage travelled by vehicles occurs in cities and large towns, although the mileage of roads in these areas is only 9% of the total in the country. However, the effect of motor traffic on small rural villages is often devastating and poses problems which cannot be ignored.

The distinction between the two situations is that large urban areas create their own traffic, whereas small villages scattered throughout the countryside have to suffer traffic generated elsewhere.

The major problems arising from the interaction between transport and the built environment are: death and injury, noise, atmospheric pollution, waste of transport resources through congestion, and visual intrusion. These problems will be examined in more detail. At this stage some general remarks on public attitudes might be useful.

Until recently, the major public concern in the matter of city traffic was with road casualties. Curiously, in recent years, documents and speeches issued by groups whose major object is the protection of the environment scarcely mention this aspect, and seem to accord much more importance to the problems of noise, pollution and visual intrusion. Over twenty people are killed each day on the roads of Britain, and over 950 are injured. Apart from immeasurable human suffering, the annual cost to the country, including loss of life, medical treatment, loss of output, administrative costs and damage to property, is estimated at over £615 million. Twice as many fatal and serious accidents occur in built-up areas as in non-built-up areas and, as might be expected, slightly over 90% of casualties to pedestrians occur in the former. Education in road safety has had a marked contribution to reducing accidents, but all the evidence suggests that the most immediate way to reduce accidents is by road improvments such as the separation of pedestrians from vehicles, and traffic control. Implementation of these measures will have as great an impact on the built environment as would the measures taken purely to improve the efficiency of traffic movement.

The built environment might be residential, or a mixture of residential and commercial in the inner areas of older towns, or commercial in town centres—each of these situations having its own peculiar problems. Conflict inevitably arises between the various interests and objectives.

All transport systems, with the possible exception of horse-drawn barges and pedal cycles, generate noise. Even pedestrians in a thoroughfare reserved for their exclusive use can create a level of noise. The channels along which noise is generated by aircraft and railways occur over a smaller proportion of most city areas than the channels along which noise from wheeled vehicles is evident. Thus traffic noise is a more general problem.

Problems of definition arise when discussing the subject of noise. In the context of traffic, the best definition which may be offered is *unwanted sound*. The intensity of sound is not the only factor involved; the duration of sound and the normal steady background are also important. One vehicle passing along a quiet residential street in the middle of the night may be more devastating to the residents than the continuous roar of traffic on a road through an industrial estate to workers in noisy factories.

At first glance the fundamental cure is to remove the source of noise. However, it may be sufficient to protect the recipient from noise in his home or work place.

In Britain and Northern Europe, there is no conclusive evidence that fumes from road traffic cause a health hazard. In some parts of the world—Los Angeles, for example—atmospheric conditions result in the formation of photochemical smog which is injurious to man and to plant life. Fumes from traffic disperse fairly rapidly in open situations, but in streets carrying heavy traffic, especially when they are surrounded by buildings or near junctions, the fumes may be concentrated to a degree which could be injurious to the health of people compelled to live or work nearby. While highway design in relation to buildings might make some contribution to a reduction of the intensity of pollution at particular places, the total amount of pollutants entering the atmosphere is obviously unaffected. Thus, the treatment of this particular nuisance differs from the treatment of noise in that an effective control must be applied at the source. Measures to this end are gradually being applied by governments throughout the world.

Another form of atmospheric pollution is that of dust. However, the amount of dust actually created by traffic is comparatively small, and the major problem is that of increased traffic creating a greater circulation of dust already in existence.

Delays in congested city streets result in a waste of resources. This is generally measured by transport planners in monetary terms, but consideration should be given to physical losses (e.g. of fuel) which are important from the point of view of long-term conservation of scarce material resources. The transport planner deals with the estimation of losses of time (especially working time) and fuel by comparing the existing situation with one which can be achieved by a proposed improvement. He is thus estimating whether the proposed saving represents a satisfactory economic return on the capital to be invested. This does not involve an absolute definition of the *congestion*. Indeed, there can be no absolute definition. The driver who is unable to make his journey at the speed he desires will consider his route congested. Even when traffic is moving at an "acceptable" speed to all drivers in a particular street, a pedestrian attempting to cross the road and being delayed in doing so might consider the street congested.

An alternative to highway improvements might be the transfer of many journeys by private car to mass transport facilities. This is a major aspect to be investigated later in the chapter. Even if such a transfer could be made on a complete scale, the impact of urban highway transport on the urban environment would not disappear.

People become so conditioned to their normal daily surroundings that the visual effect of road and rail systems on the built environment is scarcely noticed, except by those particularly sensitive to the aesthetic quality of their surroundings. The intrusion of the motor-car is most noticed by people when they are away from their normal background. Visiting other cities, they may find it impossible to photograph a street scene or a notable building without having a sea of car roofs occupying half the picture frame. The stationary vehicle, whether by the kerbside or in an open car park, is the worst offender.

The intrusion of overground railways in the nineteenth century has generally been to the detriment of the architectural background in cities, although the area affected is less than the area affected by roads.

The intrusion of waterways into cities generally shows the opposite effect from most other transport facilities, in that it is often the waterway itself which is aesthetically offensive. Many towns have recognized that by cleaning up the dereliction along rivers and canals, a satisfying visual and recreational amenity can be created.

Objectives and policies

The foregoing is a brief and perhaps superficial summary of the effect of transport on the built environment, and in general it is a restatement of commonly-accepted disadvantages. "Solutions" which have been offered range from the extreme laissez-faire (allowing the forces of congestion and cost to regulate movement of people and goods) to the complete restraint of vehicle movement (at least in central areas of cities and towns). However, it is important to appreciate that there is no real solution. There can only be a balancing of conflicting forces to achieve objectives formulated by the community. The objectives and aspirations for movement are likely to conflict with the objectives which people have for choice of a home, and both of these may conflict with other requirements for the infrastructure of a town. Indeed, the term "built environment" does not define one single environment for everyone. The built environment of the home varies from the low-density opulent suburb to the crowded dingy streets of decaying nineteenth-century housing, and the environment of the work place may vary from the factory to the office or warehouse or shop. In all discussions we must define whose environment we are considering.

However, one fact emerges clearly; we are dealing with limitations on freedom, and there can be no absolute freedom of movement if we chose to base society on urban living.

The anatomy of urban traffic

The categories of urban movement

No single measure proposed for the alleviation of transport problems in towns can possibly be applied uniformly to all the complex movements which arise. To clarify the complexities it is necessary to categorize transport movements so that individual elements can be examined. First of all, transport movements may be looked at from the point of view of the origins of the movements and their destinations. Figure 3.1 shows a simple classification for most towns and cities. The town in the diagram is reduced to very simple elements, namely, the central business core, the suburbs, and an outer boundary of the built-up area, which for our purposes need not be the administrative boundary. Large conurbations in Britain are aggregates formed by the amalgamation of smaller towns which were gradually engulfed by the expansion of cities over the last hundred years or so. These

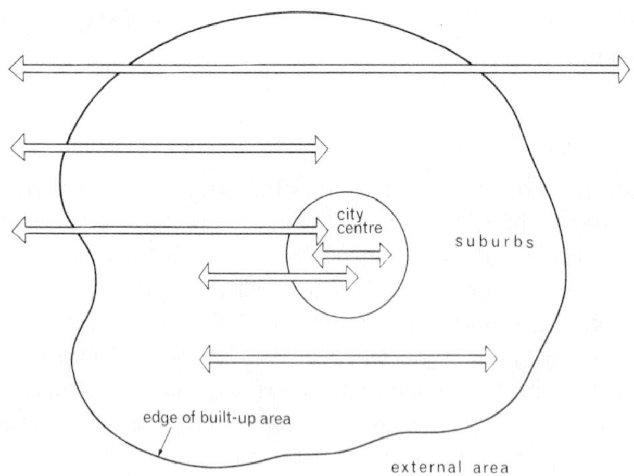

Figure 3.1 Origins and destinations of transport movements

conurbations have a major regional centre surrounded by other areas—something which is overlooked by some commentators who talk as though large cities had only one core. The seven conurbations in Great Britain contain about one-third of the population and, although they provide the most spectactular examples of the effect of transport on the environment, the problems of smaller towns in which the bulk of the population lives are of equal importance. It is, therefore justifiable to use the simple diagram shown, and the complexities of the conurbations can be constructed by the amalgamation of several such simple diagrams, with the regional centre superimposed.

Thus, origins of traffic movements can be classified broadly as follows:

Travel between points external to the town and other points external to the town ("external to external").
Travel between points external to the town and points in the suburbs ("external to suburbs").
Travel between points external to the town and points in the town centre ("external to centre").
Travel between points in the suburbs and other points in the suburbs ("intra-suburban").
Travel between points in the suburbs and points in the town centre ("suburbs to centre").
Travel between points in the town centre and other points in the town centre ("intra-central").

The movements classified so far are two-way movements; no distinction has been made between a journey beginning outside the town and ending in the town, and a journey in the opposite direction. Over a day, these two directional movements generally balance, although they do not always balance on a given route which constitutes part of a network offering alternative paths between origin and destination. Over the peak period, however, there is generally an imbalance between the number of people travelling in opposite directions. This imbalance is of considerable importance to the public transport operator, whether on rail or bus systems,and is important for the controller of traffic on the highway. It is also of importance when planning new systems. For simplicity, however, the directional nature of the movements will be ignored.

In the case of travel by car, a distinction must be made between the journey made by an individual and those observed objectively for analysis by a transport planner. For instance, a motorist may leave home, take his child to school, and then proceed to his work place. To the motorist this is really one journey, but in terms of classifying movements according to origin, destination, purpose of journey, and use of the road network, two journeys have to be identified—one associated with educational activities, and one associated with the work place. At the other extreme, a delivery van may complete one circuit of deliveries, but each "drop" has direct relevance to the activity being generated by the distribution of goods.

Besides the classification just made, based on origins and destinations, journeys by people can be classified according to the purpose being served:

Journeys to work
Journeys for business
Journeys for shopping
Journeys for education
Journeys for personal business
Journeys for social purposes.

The first category can be further sub-divided according to the category of work. This could be of importance when examining the relation between economic activity and movement, e.g. the manual industrial worker will have a destination generally outside the city centre, whereas white-collar workers engaged in commercial and service occupations will more likely be travelling to the business centre of the city. Similarly, travellers to educational establishments might be subdivided into students, teachers, administrative and technical staffs, and maintenance workers. The term *social purpose* can include journeys for medical and dental services, visits to hospitals, lawyers, bankers, insurance or other service agencies, sporting or cultural centres.

If the actual statistics of travel for a town were to be presented, they might appear in the form shown in Table 3.1. This table shows 36

Table 3.1 Classification of journeys by origin and purpose

	Purpose of journey					
Origins and Destinations	*Work*	*Business*	*Shopping*	*Education*	*Personal business*	*Social*
External to external						
External to suburbs						
External to centre						
Intra-suburban						
Suburbs to centre						
Intra-central						

categories. This is almost the minimum number which can be used if we are to relate, as we must do, the journeys to the activities which generate movements.

The means which people use to make journeys are car, bus, rail, cycle and walking. Waterborne transport has been omitted, although in some cities it is available for some part, if not all, of the journeys. Some towns do not have rail transport available for suburban transport, but it may be available for journeys to and from points beyond the town. Air travel is not included, as landing generally takes place outside built-up areas, and the passenger completes his journey on the ground. This situation could possibly alter in future with the development of VTOL (vertical take-off and landing) aircraft.

Many journeys consist of combinations of means of travel, e.g. a car may be used to reach a railway station, and the end of the journey then completed by walking or using a bus. However, if we keep to the five categories listed above we have to muliply the 36 categories in Table 3.1 by five to obtain a simple classification of these journeys. The resulting number of categories is 180.

Turning to goods traffic, the same classification for origins and destinations may be maintained, and journey purposes classified as:

Movement of raw materials.
Movement of components and semi-finished products.
Movement of finished products to distributor or user.
Distribution of retail products.

Table 3.2 shows how statistics of goods movements in a town might be presented, and it will be seen that there are 24 categories. These movements may be made by road, rail, air or waterborne transport.

In general, where the main element of a journey is by rail, air or waterborne transport, one part of the journey is likely to be by road—except for certain bulk products carried by the "merry-go-round" system. However, as far as town traffic is concerned in Britain, rail, air and waterborne systems are unlikely to be used for movements within the built-up area, and distribution and collection of goods being carried by these systems is likely to be by road.

As the journeys with an external origin or destination may consist of movement from the town by rail or water, these categories must at least be maintained. It is convenient, therefore, to classify the means of movement into road, road and rail, road and water, road and air.

Taking these four categories into account, in addition to the 24 already recognized for goods traffic, 96 classes are obtained.

Table 3.2 Classification of goods movements

	Purpose of journey			
Origins and Destinations	*Raw materials*	*Components and semi-finished goods*	*Finished products*	*Retail distribution*
External to external				
External to suburbs				
External to centre				
Intra-suburban				
Suburbs to centre				
Intra-central				

The categories of land use

The classification of origins and destinations shown in figure 3.1 is extremely broad. Even if traffic movements were generated uniformly over the three defined areas—external, suburban and central—it is clear that any given category of journeys classified by purpose is unlikely to be uniformly distributed. Certainly goods traffic will not be uniform throughout the area. For a more complete analysis of urban movements, the origin and destination of each class of movement must be related to the location of the activity which generates the movement concerned. Land use can therefore be classified in accordance with the social and economic activity which characterizes the particular location. The simplest classification might be:

Residential
Manufacturing industry
Service industry
Retail shops
Wholesale distribution
Offices
Public buildings

These categories are obviously very broad in application. Transport facilities themselves, in the form of railway and bus stations and ports, not only concentrate movements through them at a particular location but constitute centres of employment. The differing characteristics of manufacturing industries in terms of employment intensity will have an effect on the movements generated. However, sufficient has been said to show the relation between journey purpose and the location of the activity where the purpose will be fulfilled.

When the transport planner is conducting an analysis of traffic movements in a town, he divides the town into zones, each of which represents a "catchment" area of traffic generation. About one hundred such zones might be used for a town, although these zones might be subdivided for some purposes of a study and aggregated for others. In the case of one hundred zones, it will be seen that there would be 100×99 possible movements between zones. Naturally, these movements are consolidated into a much smaller number of paths through the area depending on the network available. Thus any of the 9900 movements between zones could in theory be composed of a possible 36 categories from Table 3.1, multiplied by five categories of means of transport, all associated with seven classes of land use. For goods traffic there is a corresponding complexity. Obviously, not every interzonal group of movements will contain all the classes mentioned; if there is no rail connection between any two zones, movement by that form of transport will not exist.

Some quantitative considerations

The formulation of policies and, in particular, the comparison of alternatives must depend on measured data. The figures and facts must at some stage be related to the costs of implementation of a policy. All policies must attempt to deal with future transport demands, as the physical results of investment must serve the community for a long time. Towns vary in their activities, physical structure and transport systems, so that each case must be considered on its merits. One can do no more than indicate the range of values for various traffic elements which might be encountered in present-day cities. Some of these indications might assist in enabling us to examine in general terms contemporary proposals for policies to meet the needs of traffic in towns.

Town transport movements previously defined as *external to external*—clearly do not serve any purpose—social or economic—of benefit to the town. Movements by water and air in this category are irrelevant as a rule. Movement by rail often combines other categories of movement, e.g. *external to centre* in the same train over the same track, so that the elimination of extraneous *external to external* traffic would not result in the saving of any provision of facilities.

Road transport is generally in a different category. Figure 3.2 shows that this category of traffic varies with the size of the town concerned. This figure further demonstrates that the impact of through non-stopping traffic on small towns and villages is very serious. Conversely, the elimination of non-stopping through traffic from larger towns and cities has a relatively small effect, and such elimination can only be made by the construction of very long by-passes.

Clearly, large cities generate their own traffic, and, unless growth of cities is checked, increasing urbanization throughout the world will result in even more movement within cities in proportion to population. It must be borne in mind that figure 3.2 relates only to vehicle movements. When account is taken of public transport, movements of people are, of course, greater. However, this does not affect the main issue.

It is not easy to derive from a study of various traffic reports how many

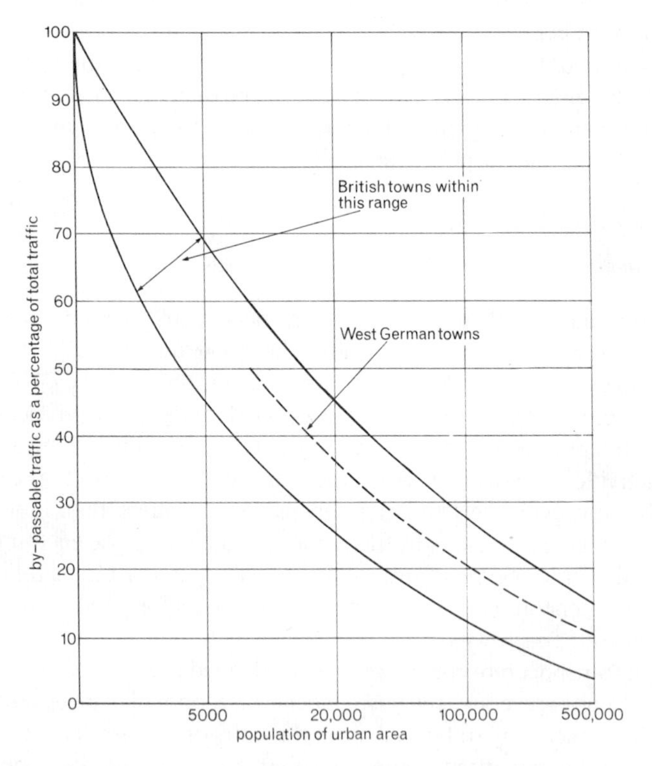

Figure 3.2 Road traffic and size of town (compiled from *Traffic in Towns* and *Conference on Urban Motorways*)

people travel *external to suburbs* and *external to centre*. It is much easier to establish the figures in terms of private cars. As we are primarily concerned with the proportions and relative importance of different movements, figures in term of private cars are useful indications. For large cities about 25–35% of all car traffic entering or crossing the city is *external to suburban* traffic, and about 15–20% is *external to centre*. It is worth noting that in terms of travel by persons, a large conurbation shows that about 40% travel from external points by car, about 30% by bus and about 30% by train. Most of the people travelling by bus or train have destinations in the city centre, some have destinations at points along the routes served by bus and train (which are primarily directed to the centre), and the smallest category travel to suburban points. On the other hand, the proportion of car travellers (including passengers) with destinations in suburban zones is relatively higher than for public-transport passengers. These results are not unexpected but it is surprising that the implications of these facts are often ignored by those who advocate policies of changing from car travel to public transport at some point in a journey. This will be discussed later, but for the present it is sufficient to note the large element of car travel from outside a city to a diffused pattern of destinations in the suburbs.

Let us now consider the movements which occur within a city i.e. from *suburb to suburb* and from *suburb to centre*. As figure 3.2 shows, these two categories constitute most of the movements in larger towns and cities. By far the largest category is the journey to work, which constitutes about 40% of all travel on an average working day. The relative proportions of travel from *suburb to suburb* and from *suburb to city centre* obviously depends on the distribution of work places.

In Britain, for towns with a population of over 100 000, about one third to one half of workers travel to the city centre. Most workers in the city centre travel by public transport, the proportion varying from about 50% for smaller towns to 80% for large cities (especially those which constitute the core of conurbations).

For travel between points in the suburbs, the proportion of travel by private car is much greater than in the case of travel to the city centre. This is not unexpected, as most public-transport systems are directed towards the centre and, even where circumferential services exist, they serve a fairly narrow band of population and work places.

Naturally, most work journeys create the familiar problem of "peak" travel. Each morning the radial routes to city centres are congested with this peak-hour traffic, and many observers have the impression that this traffic is flowing into the city centre to work. If we consider vehicles instead of persons travelling, we obtain a much modified view. Surveys have disclosed that in some of our larger cities 35–40% of vehicular traffic entering the central business district during the peak hour does not stop in the centre, but consists of *suburban to suburban* movements; in some cases the stream includes *external to external* traffic, although, as has been pointed out, the proportion of such traffic is small in larger cities. In one city the proportion of traffic *crossing* the central business district is as high as two thirds of all traffic *entering* the central business district.

The number of people working in city centres, both in Britain and in North America is declining. In the period 1961–1966, there was a fall of 8% in traffic entering central London. Peak travel has remained constant in some cities in North America, although the population of the cities has grown. In Britain, many cities have seen a decline in population. This decline has been a result of deliberate policy in some cases, especially in association with the construction of new towns. Some of the workers who have moved to new towns continue to work in the city from which they have moved. However, on balance, there has been a decentralization of both population and work places. This trend seems to create an anomaly, in that the physical provision of offices and shops in city centres seems to be growing in many cities, although travel to work in the city centre has remained relatively stable. The explanation lies in the provision of greater office space per person. In the case of retail trading, greater space is provided per person working but the nature of the trade has reduced the number of people working in relation to the space provided. The trend to greater space per person must slow down, if it has not already done so. However, the danger of assuming that movements, especially by private car, will grow in city centres at the same rate as throughout the country generally is emphasized.

The decentralization of activities from city centres not only affects traffic moving in and out of the centres but also increases the number of journeys from *suburb to suburb*. These journeys are widely dispersed and constitute a type of movement not easily tackled economically by public transport, which requires considerable consolidation of movements.

In discussing movements of people, a distinction has been made between numbers of travellers and numbers of vehicles in which they travel. It is difficult in the case of goods vehicles to make an analogous distinction between goods vehicles and the goods which they distribute. Research is continuing in this difficult area, but for the present we shall consider movements of vehicles only. Goods vehicles constitute about 12% of the "average" traffic" stream in urban areas, although there are considerable variations about the average, depending on whether the road concerned is an industrial or dock area, or in a suburb. There are also variations with time of day. When comparison is made with North American cities, where car ownership is near "saturation", it would seem that the proportion of commercial vehicles will stabilize at about 9% of the total traffic—except, of course, in situations where there is a prohibition on the use of a street by particular classes of vehicle.

The distribution of goods traffic between the categories of origin and destination defined in Table 3.2 follows surprisingly closely the distribution of light vehicles between these categories. The problems of goods traffic extraneous to a town and to the business centre of a town area are, in relative terms, the same as for private cars, except that the intrusion of heavier and noiser vehicles might be considered to have a worse effect on the built environment than the intrusion of private cars. It is worth noting that, because of the increasing carrying capacity of goods vehicles, the growth of commercial vehicles is not proportionate to the growth of production and movement of goods.

Most of the quantitative considerations discussed above have been in relative terms, i.e. proportions have been used in place of absolute numerical values. The professional transport planner must, however, deal with the different categories and their interaction in absolute terms. These interactions are expressed mathematically to create models which can be used for a quantitative analysis of the consequences of a particular proposal.

Impact of urban transport on the built environment

General considerations

The major impact of transport has already been described as death and injury, noise, pollution, congestion and visual intrusion. Now that the traffic in urban areas has been classified, it is possible to consider the contribution that each category of traffic makes to the total effect of traffic on those who have to live and work in cities. A closer examination is, therefore, necessary and, in addition to the examination of the anatomy of urban traffic made above, some consideration must be given to the anatomy of the physical nature of highway systems.

The highway pattern

Most literature dealing with highways and town planning classifies highway networks in urban areas into groups: grid-iron, radial, linear or irregular. It was sometimes claimed that these patterns were apparent from a study of plans of existing cities.

If reports of modern traffic studies or plans of cities prepared by motoring organizations to show through routes and congested routes are studied, it will be found difficult to classify highway patterns if useage by traffic is considered instead of purely geometric pattern. This difficulty was recognized at an early stage by transport planners who realized that the adoption of idealized patterns took little quantitative account of demand for movement in relation to motive or purpose; thus the first attempts to relate movement to land use began.

Studies of this kind confirmed by measured data what is apparent to the ordinary observer, namely that the road patterns of our cities bear little relation to modern needs, apart from their obvious lack of capacity to deal with contemporary volumes of traffic. The general pattern of most cities shows a convergence of arterial roads to the centre of the city. In the past these roads were used only by the traffic defined above as *external to external* and *external to centre*. There were no suburbs in the sense understood today. Even after the Industrial Revolution, industry located in cities was within walking distance of workers' homes. Later, with development of tramcars, longer journeys were made to work, but an examination of old maps of industrial cities shows that mills, factories, and shipyards were located along the tramcar routes quite as evenly as homes. In spite of redevelopment, many cities still show the remnants of old factories, often disused, in the central area. They will disappear in due course but, except when redevelopment changes the road pattern, the old transport system will remain. The wealthy lived in the suburbs, but the housing reforms which led to workers living in suburbs began primarily at the beginning of this century, and the movement reached greatest significance between the two great wars.

Location of industry changed, and the creation of "industrial estates" began between the wars. These estates were created in response to changes in economic and social factors, and changes in the nature of production. Whatever the reasons, until recently the major road patterns of cities remained as they had been before the pre-industrial age or in the early stages of that age.

If we go back to the classification of journeys which was made in Tables 3.1 and 3.2, and apply these categories to a road pattern which grew up in an entirely different age, we can see more clearly the real nature of the problem which is causing all thinking people so much concern today.

The evolution of most industrial cities resulted in the central business area expanding, with shops and other commercial premises along the arterial roads entering the centre. As distance from the centre increases, there is a change to residential property, interspersed with occasional minor shopping centres. In some cities factories are sited along arterial roads at some distance from the centre.

Traffic entering a city area from the surrounding rural area will consist of *external to external* traffic, together with *external to suburbs* and *external to centre* traffic. The amount of the last two categories will depend on the extent of the links between the city and other cities or towns, and on the extent to which people who work in the city chose to live in other parts of the region. This latter class has increased in recent years, partly because of increasing affluence, but also because of a transfer of people from rail travel, which in the past enabled quite substantial numbers of people to travel from points well outside the town's immediate environs.

The first section of arterial road from the boundary will generally pass through a residental area, and the impact on this area is generally sufficiently serious to destroy all the advantages of amenity which the original settlers in these areas sought.

From the boundary inwards, the arterial road serves as a collector of traffic from the areas between the radial routes. Some of this traffic will be bound for the business centre, but some will continue through the centre to destinations beyond the centre. In some towns it is possible for such traffic to follow circumferential routes, but in many cities such routes are so circuitous and cut across so many radial routes as to be unattractive. Besides, circular routes of this sort often use residential roads which were not originally intended to carry this type of traffic. There comes a point close to the outer edge of the central business district where the volume of traffic is a maximum, and at this point it is passing through areas of shops and houses which spread from the centre.

Within the central area, traffic (other than through traffic) begins to spread out through the network to individual destinations conflicting with other similar movements contributed by other radials. Over the length where traffic has reached its maximum volume on the edge of the central business district, there occurs the maximum conflict with the environment. Not only is there the noise nuisance at a maximum, but the lack of capacity of the road results in measures to ease the flow of traffic, consisting of prohibition of parking, loading and unloading, making turns in specified directions. The shops along such routes lose custom, and find trading inconvenient, so there is a movement of the better businesses to other sites, and poorer-quality businesses move in; property values fall. The deterioration due to these changes added to the more direct effect of traffic causes a similar fall in values for residents, and again a social change follows. One serious effect is the disruption of communication between residents, especially older residents, which takes place. Residents on one side of a street which carries arterial traffic rarely meet those on the other side, and indeed the general change in character results in a reduction in social contacts which could be described as a loss in "neighbourliness". In one city, some years ago at the time of a general election, the agents for all political parties protested that there were no polling stations in an area bounded by a busy arterial road and a river. The agents asserted that some of the older people had not left the area for many years, because they were afraid to cross the busy road, and they would be deprived of their vote because of this quite reasonable fear for their safety.

Accidents tend to increase as traffic volume increases, though severity of injury and damage tends to fall as a result of lower traffic speeds. Nevertheless, movement of traffic in arterial streets tends to be fairly continuous, so that pedestrians have fewer opportunities to cross than are available in the central business district itself, where more frequent intersections provide more crossing opportunities.

In the central business district, the impact of traffic on residents is not generally the problem, although this clearly does not apply in smaller towns and villages where there are proportionately more residents in the centre.

The impact of traffic in the centre on business is severe, and it must be emphasized again that a high proportion of the traffic which causes the deterioration in business activity has in fact no business in the area at all. These remarks apply to circumstances prevailing before the construction of relief roads—which has taken place in many cities. It is necessary to point what the situation was before the construction of such roads, in order that we may assess later whether such relief routes constitute a satisfactory contribution to the amelioration of the effects of traffic on the built environment.

Safety

In the introduction to this chapter some general figures were given to show the scale of accidents on the highways of Britain.

It is now necessary to examine the situation more closely in the context of the analysis which has been made of different types of movement and different types of road. A difficulty immediately arises if statistics are to be used to make comparison.

It is of little use stating that circumstances are more dangerous on one road than on another, because the number of accidents on the former is greater than on the latter. The road with the higher number of accidents may be three miles long and carrying a large volume of fast traffic, whereas the road with the lower number may be half a mile long with only two or three vehicles using it per day. Some attempt has to be made to bring figures to a common base before making comparisons. In the case just quoted it may be that, after the figures have been corrected to some common base, the apparently quieter road turns out to offer a higher risk of accident. It does not necessarily follow that, if resources are limited, it is better to take measures to improve the quieter road with the higher risk. The decision may well be to reduce the absolute number by measures applied to the busier road, especially if the amount of investment is relatively less in relation to the benefit to be achieved.

The common base which may be used for comparison is expressed in the form of a *rate*. There are many ways in which accident figures can be expressed as a rate but, before proceeding to formulate these rates, the nature of the accident figures themselves must be clarified. Some countries use statistics based on all reported accidents. Figures expressed in this manner cannot be compared with figures in the United Kingdom, because in this country all accidents are not reported to the police nor need they be reported when personal injury or damage to property is not involved (provided certain formalities are observed). In the United Kingdom, therefore, comparisons are based on accidents involving personal injury. However, the number of such accidents on a length of road may not be the same as the number of casualities on a road, as several people may be injured in one accident. Injuries may be classified as "slight", "serious" or "fatal", and the ratio of serious and fatal accidents to the total is an indication of severity. This leads to a further complication being presented to an authority concerned with the reduction of accidents. There may be fewer accidents on one road than on another, but they may be of greater severity, involving more loss of life.

Accidents may be further subdivided into categories such as car drivers, motor cyclists, pedal cyclists, pedestrians, bus passengers, and age categories.

Whatever class of accident is being investigated, comparisons must be made on a common basis, i.e. "rates" must be compared rather than absolute numbers (with the proviso already mentioned above that policy decisions may be influenced by numbers rather than rates in particular circumstances). There are several rates which may be used, and in any discussion the particular rate being quoted must be specified.

For the sake of brevity the term *accident* will be used in what follows, but it must always be remembered that this term must be specified according to the categories given above. Moreover, we must specify an area (e.g. an administrative area, a length of road, a junction or whatever the location is) to which the figures apply.

Rates which may be used are:

accidents per annum (or other period of time)
accidents per vehicle (class to be specified)
accidents per driver (class of vehicle to be specified)
accidents per head of population in the area concerned
accidents per unit length of road
accidents per unit of flow on a road (e.g. accidents per thousand vehicles entering an intersection per day).
accidents per million vehicle-miles on a specified length of road (kilometres now in general use).

It is astonishing how frequently references are seen in the columns of the press to an "accident rate" without the rate being specified. This results in most misleading statements being made.

It would be interesting and highly relevant to the question of the impact of traffic on the built environment to elaborate on the subject of safety but, for the purpose of examining how policies of transport reform in cities may relate to safety, it is sufficient to use an accident rate which makes a basis for comparison between different categories of road. The rate used here is the number of personal injury accidents per million vehicle-miles. There is, of course, some variation within each category, but the figures quoted are average figures derived from several sources. The following list gives these rates, and a column is appended showing each rate as a ratio of the rate for motorways. This is an indication of the relative risk of personal-injury accidents occurring, but it is not an indication of the relative risk to an individual road user.

Table 3.3 Accident rate

Category of road	*Personal-injury accidents per million vehicle-miles*	*Relative accident rates*
	(All figures rounded off to nearest whole number)	
Urban all-purpose roads		
Shopping centres	11	18
Residential roads	7	12
Major busy roads	4	7
Dual carriageways	5	8
Dual carriageways with grade separation at major intersection	2	3
Rural all-purpose roads		
Minor roads with light traffic	2	3
Major roads with heavy traffic	3	5
Dual carriageways	1·6	3
Special roads (motorways)		
(No distinction made between urban and rural)	0·6	1

The professional transport engineer would not use terms like *heavy* and *light* traffic. He would specify conditions within ranges of volume of traffic, but the simplification involved in the table by this and other considerations does not affect the general order of magnitude of the figures given. The table makes no distinction between serious (including fatal) and slight injuries. If such a distinction is made, it is found that the proportion of serious accidents on a major road passing through the suburbs is much higher than on equally busy major roads in the business centre of a town.

An example of how these rates might be applied can be given by considering a length of one mile of busy shopping street and assuming that a motorway (or similar facility) is constructed of the same length to cater for non-stopping through traffic. If V is the volume of traffic (in millions of vehicles) before the construction of the by-pass, then the number of accidents likely to occur over the period for which the volume of traffic has been measured will be $v \times 1$ mile $\times$ 11, that is, $11V$, using the accident rate given in Table 3.3.

An assumption that 35% of the traffic in the street is by-passable is not unreasonable considering figures already given for this category. After the construction of the motorway we have:

Personal injury (P.I.) accidents on motorway $0{\cdot}35\ V \times 1$ mile $\times\ 0{\cdot}6 = 0{\cdot}21\ V$
P.I. accidents on shopping street $0{\cdot}65\ V \times 1$ mile $\times\ 11 = 7{\cdot}15V$ which makes a total of $7{\cdot}36V$ accidents.

The reduction in accidents is 33%.

If a decision was made that no wheeled traffic was to be permitted in the shopping street and that all vehicles would halt in adjacent car-parks, leaving the street for the exclusive use of pedestrians, then the reduction in personal injury accidents would be 11 minus 0·21 or a reduction of 98%.

Except in the case of the simplest problem of a by-pass to a small town built along a single arterial road, the real-life situation is more complex than the example implies. In an urban network the re-distribution of movements consequent on major construction is much more complex, and consideration has also to be given to changes in the composition of traffic and changes in speed which affect the severity of accidents. Nevertheless, the example indicates the general use of accident rates and gives some indication of the effect of changes. If we turn the problem the other way round, and start from a situation with no through traffic in a shopping street, or none at all, then we can interpret the difference in accidents as the penalty which is paid for allowing wheeled traffic into areas and over roads not designed for modern conditions. This is the situation which was examined above in the analysis of conditions on an arterial road entering a town.

It could be argued that this impact of traffic on safety is not an impact on the built environment, and indeed some authoritative official papers dealing with the subject of the environment do not include safety in their studies. This is to take too narrow a view.

The built environment is where people live and work, and movement in and out of homes to shops, schools and work places is conditioned by the nature of the built environment (especially the density of building) in relation to accessibility to the facilities which people must use in their daily lives. Besides, if consideration of policies is restricted only to other effects than safety, then some proposal might improve the built environment at the expense of more injury on the road. Alternatively, the question might have to be considered whether some standard of built environment short of perfection might have to be accepted in exchange for a saving of life and limb.

Noise

The World Health Organization (WHO) defines health as "a state of complete physical, mental, and social well-being, and not merely an absence of disease and infirmity".

It appears that the general effect of noise on health is more psychological than physical, though there are specific cases where noise has a physical effect, and not only can pain be caused but damage may be sustained to hearing. Such cases are rare in relation to road traffic noise, except for situations where a driver may be affected. Such a case would, however, relate to the individual vehicle and not to traffic noise generated by total traffic on a roadway over a period of time.

Interpreting the WHO definition of noise in its widest sense, it might be said that road traffic noise has an effect on health in that such noise reduces mental and social well-being.

There are problems of noise generated by other forms of transport, the most obvious being aircraft. A survey in London disclosed that railway noise predominated at only 4% of sites examined, whereas road noise predominated at 84% of the sites.

In a survey more people complained of disturbance from road traffic noise when at home than they did when at work. A higher standard of peace and quiet is demanded at home than at work (generally), especially when the activity at home may be listening to music or studying. The actual physical measurement of sound pressure level does not always indicate the degree of nuisance created. The distribution of frequencies within propagated sounds may vary for the same degree of "loudness", and certain frequencies may be less acceptable to the recipent than others. The effect of traffic noise may be to create difficulty in communicating by speech with others, to reduce efficiency at certain types of work, and to interfere with sleep. Research has related the general dissatisfaction arising from noise as expressed by subjects in varying situations to physical measurements made by apparatus.

Unfortunately, there is no single apparatus suitable for measuring noise level for all types of noise sources. The apparatus used to measure road-traffic noise levels gives a reading in decibels, the machine being fitted with a frequency weighting filter in an effort to obtain readings which correlate with the response of the human ear to traffic noise. The particular weighting used for road-traffic noise level is known as A and the unit therefore used for measuring road-traffic noise levels in the 'A' weighted decibel indicated by dB(A). Measurements in dB(A) units are not directly proportional, e.g. the noise level of a soft whisper measured 5 ft from the source is 34 dB(A), and the noise level of a heavy diesel lorry measured 25 ft from the source is 92 dB(A), which is roughly three times the measurement for a whisper. However, using a subjective assessment of loudness it is found that the diesel lorry is 70 times louder than the whisper.

For purposes of comparison beweeen one road and another or between one condition on a road and another (e.g. when traffic volumes are increased on a particular road) the noise level which is exceeded for 10% of the time interval between 6 a.m. and midnight is used. Government regulations in Great Britain specify standard conditions for the way in which measurements are to be taken or in which calculations for noise levels are to be made. These procedures must be followed in order to determine entitlement under Noise Insulation Regulations 1975, and, in addition, used in calculations by highway engineers when studying the environmental impact of proposed new or improved roads.

Under the regulations, residential property may be eligible for insulation if three conditions are satisfied at a point one metre from the facade of the building involved. The three conditions are:

(1) that within fifteen years from the date of opening to public traffic of a new or altered highway the noise will reach at least 68 dB(A) for more than 10% of the time between 6 a.m. and midnight,
(2) that traffic noise within the 15-year period will be at least 1·0 dB(A) higher than before work began on the new or altered highway,
(3) that when the noise from the use of the new or altered highway is added to noise from other highways in the vicinity, the total noise is increased by at least 1·0 dB(A).

Many factors affect noise level as recorded at the reception point. For a level road carrying private-car traffic only, the basic noise level at a reception point 10 m from the edge of the carriageway varies from 58 dB(A) for 1000 vehicles in an 18-hour day to 78 dB(A) for 100 000 vehicles in an 18-hour day if the average speed is 75 km/h. The noise level increases as the proportion of heavy vehicles increases, and also as the average speed increases. These corrections may counterbalance in situations where an increase in heavy vehicles and in total volume produces a decrease in speed.

Furthermore, the steeper the carriageway, the greater the effect on traffic noise. Increased distance from the edge of the carriageway reduces the sound, although the nature of the intervening ground will affect the attenuation of sound. Long barriers have also a screening effect and, of course, depressed carriageways result in screening, whereas elevated carriageways have a greater effect on tall buildings, though buildings at ground level may be screened to some extent.

For all of these factors there are techniques to enable the highway engineer to predict the effect of a proposed highway.

The significant advance which has been made in recent years is that it is now possible to compare the amount of money which might be required for sound insulation with the cost of road design measures which would reduce noise levels. This goes a long way to enabling hitherto unquantifiable environmental factors to be assessed on a rational basis.

Mention has been made of the fact that the dB(A) scale is not uniform. One effect of this is that the addition to a noise of one other equal noise source increases the noise level by 3 dB(A).

Taking an urban area as a whole, the environment can be greatly improved by concentrating traffic on a few main routes. Naturally, the consequent effect on the immediate neighbourhood must be studied with care. From an economic point of view, expenditure to protect people living near the main route might be relatively less than the cost otherwise imposed by traffic on the residental routes to be replaced.

Measures which can be taken to reduce the noise nuisance may therefore be summarized as follows:

Noise may be reduced at the source, and Government regulations in many countries exist to this end.

Traffic management schemes may be employed to smooth the flow of traffic, i.e. to reduce the frequency of stoppages. Unfortunately, no satisfactory criteria have yet been developed to deal with this particular localized situation.

Speed limits may be considered to reduce noise, as well as from the point of view of safety.

New roads may be constructed to by-pass villages and small towns and to canalize traffic in large towns past areas where the environment should be protected.

Highway design measures such as depressed highways, and landscape-incorporating earth forms as sound barriers may be introduced.

Insulation of property.

Design of new buildings near major highways in such a way that noise-sensitive working or living areas are on the side of the building remote from the highway.

Atmospheric pollution

There is no convincing evidence that pollution from motor vehicles is creating a general health hazard. There are some special situations where people are exposed to a high concentration of exhaust fumes for long periods, and in such situations there is a hazard. Even if there is no health hazard in general terms, exhaust fumes are objectionable because of smell and smoke. The amount of pollutants varies according to whether engines are idling, accelerating, decelerating or cruising. Gradient of the road also affects the work imposed on an engine. Thus, hazards are likely to occur at busy junctions and are relevant only to people working and living in buildings close to the point at which vehicles stop and start. Although the only way to control atmospheric pollution from vehicles is at the vehicle itself, some amelioration of the effects can be derived by changes in the highway pattern and by traffic management, e.g. a by-pass or relief road will divert traffic away from streets where such traffic has no business and so reduce the area affected. By co-ordinated design of the new road, and the buildings nearby, people living and working along the line of the new road can be protected. If the new road is in rural surroundings, then no-one is affected, and in one case in the UK a motorway eight miles long diverted the by-passable traffic from a road which had 1898 access points. Not all of these points would handle traffic sufficient to create a serious pollution problem, but several hundred were intersections where there was a considerable volume of stopping and starting traffic in built-up areas.

Traffic management schemes, e.g. those involving restrictions on turning movement and those using area control of signals, result in a smoother flow of traffic, with a consequent reduction in air pollution.

Visual intrusion

Allusion has already been made to the effect the intrusion of motor traffic has on the visual nature of the built environment. It may be true, if unfortunate, that many people take the visual background presented by their environment for granted. There is, indeed, a possibility that people accustomed to an undistinguished, if not actually ugly, background may fail to acquire a reasonable level of taste in visual matters unless, through education, travel, or the possession of an innate sensitivity, a critical awareness is cultivated. The visual background of a town or city forms part of the "quality of life". The only cure for the bad effects of the intrusion of motor vehicles is their removal from the scene, with the proviso that in some circumstances the presence of vehicles does not necessarily offend, and indeed may be a lively part of the townscape.

If vehicles are removed, two problems arise. In the case of parked vehicles which are moved to a covered (generally multi-storey) car park, the problem is changed from one of the relation of vehicles to the background to one of relation of the buildings to the background. Not all examples of multi-storey car parks convince one that such structures can be satisfactorily integrated with adjacent buildings, though there are some multi-storey car parks which are the best pieces of architecture in the immediate area concerned.

In the case of the removal of moving vehicles from existing streets to a new road, the problem may now become one of the reconciliation of the new road structure and its "furniture" of signs and signals with its adjacent background.

These are problems which are capable of solution, and often a highway facility whether at ground level or elevated can be of interest and aesthetic value in its own right and in its relationship to its surroundings.

A common element

Certain conditions require to be satisfied to ensure that the relief of one section of the community from bad environmental effects is not at the sacrifice of the environment of other members of the community. However, if one element of the problem is improved by new construction, improvement to the others automatically follows.

Possible measures

Engineering construction

Road traffic which is extraneous to a village, town, city, or district within a town or city, may be diverted to its own channel by the construction of a new road. Figure 3.2 shows what relief may be expected according to the size of the town.

The by-passing of a small village or town is relatively simple, but the consideration of a larger town or city must take account of the differing requirements of the groups shown in Tables 3.1 and 3.2. The by-pass or ring road has always held an attraction for town planners in British thinking but, as long ago as 1949, Cardell pointed out the problems involved for a larger city. Figure 3.3 shows the general structure of industrial cities in Great Britain. There are some obvious exceptions to the general pattern illustrated, e.g. towns less affected by the industrial changes of the nineteenth century and which have maintained a central area of historical and architectural importance. These towns have often developed industry at a later stage, but the general character of the central area has been maintained. However, the general case which the diagram illustrates consists of a central commercial and business core of high value. Whatever changes have taken place over the years, this core maintains its high value and intense level of business activity, although the buildings may be renewed from time to time. Radiating from the core are strips of high value along the major roads which (unless affected in recent times by traffic problems) also maintain their value. Around the central core is a "twilight" area consisting often of housing which at one time would be of a

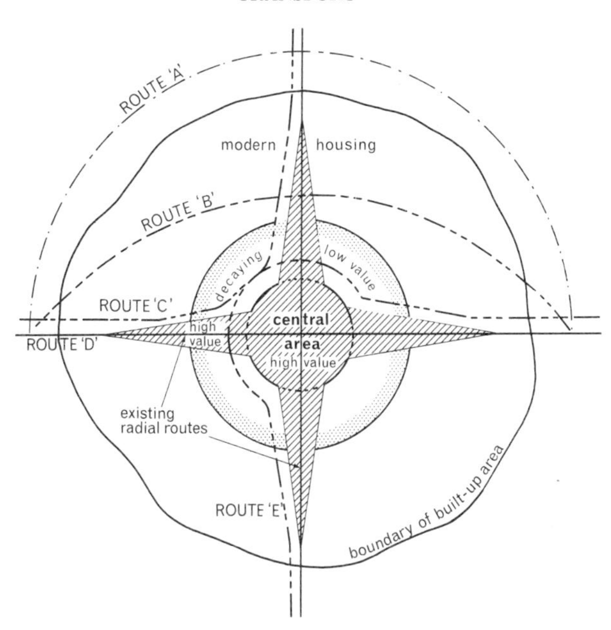

Figure 3.3

...or type, but is now decaying or turned into multiple occupancy. Sometimes this annular ring consists of nineteenth-century workers' dwellings below the standard acceptable today. The next ring of buildings consists of housing built between the wars, and the outer periphery of this ring consists generally of building since the second world war. In general, it may be said that the value of the building increases nearer the edge of the built-up area.

This outer area is interspersed with industrial areas, schools and minor shopping centres.

If we consider, first of all, the diversion of *external to external* traffic by the construction of a by-pass on line A in figure 3.3, it will be evident that any such by-pass located outside the built-up area will present a devious route unattractive to through traffic. Indeed, the diversion could be such that the advantage of higher speed to diverted traffic would be offset by the extra distance travelled. To make the by-pass less uneconomic to diverted traffic, and more likely to attract more of such traffic, the ring could be made shorter as shown by line B. Such a location would involve the destruction of comparatively newly built homes and impose disturbance on others. Some cities might present circumstances which would permit a road to be located through such an area, but the likelihood is that if a strip of land is free of building it is also unsuitable for road construction or it would mean more expense than normal. A ring road on the line B would perhaps also serve some intersuburban journeys, but this contribution is likely to be small in view of the diffuse nature of such journeys.

Neither Route A nor Route B would make any contribution to those internal movements of the city traffic which cross the central area without having any need to stop there. This category of movement is substantial and, if relief is to be given to the central complex, a second facility would be required.

Consider now a route along line C in figure 3.3. The assumption is that the routes being examined do not have frontage development with freedom of access anywhere along the line. To serve the intended purpose access must be limited, and a high degree of free flow given to traffic. It might be that motorway standard would be required. Nevertheless, it would be possible to provide a sufficient number of access points along Route C to enable a road on this alignment to serve *external to external*, *external to centre*, *suburb to suburb* and *suburb to centre* movements. The degree to which such a route serves each of these movements depends on th accessibility of the route and the distance to be travelled. It must be borne i mind that the structure of most of our cities is such that the existing radi system (Route D on figure 3.3) already serves these movements, and effect of constructing route C would be to remove all but local traffic the existing radial route. Many such existing radial routes are so cong that parallel residental streets are often used to give relief in one-wa systems. This results in deterioration of the environment, with a consequent fall in property values. The relief offered by Route C has, in actual cases, resulted in a restoration of amenity to these residential streets, with a consequent increase in property values, and indeed the conditions on the main radial have also been improved.

These benefits, together with the benefits to the central business district, are achieved at the expense of the strip along which the new road is constructed. Route C results in the least disturbance of all the alternatives which might exist between the outer limit and Route C. Route C is, in the first place, shorter than the alternatives; the damage to the outer high-value zone is at a minimum, as the route crosses the annular ring radially and not circumferentially. Incidentally, it is sometimes the case that old and decaying development lies behind the shops and businesses which line the radial, so that it does not always follow that recent relatively high-value property is affected. In one major instance the road through this zone was a result of the clearance of slums and the redevelopment of the area, not the other way round. There were pockets where the population density per acre was over 600 and where more than a dozen households shared a single cold-

combination of engineering and legislation. They consist of restrictions on waiting at the kerbside, on loading and unloading; prohibition of certain turning movements; reserved bus lanes; one-way streets; traffic signals and other measures at intersections such as islands, roundabouts and box-markings.

Where these measures reduce accidents, produce a smoother flow of traffic with fewer stoppages and thus less noise and pollution, the impact on the built environment is good, but some of these measures often produce a lowering of the quality of the environment, e.g. a one-way street system may result in the diversion of traffic through hitherto quiet residential streets. Speed of traffic may be increased, resulting in greater difficulties to pedestrians crossing the street although, in spite of increased speeds, accidents are generally reduced.

Area control of traffic signals by computer has not only resulted in higher journey speeds for traffic but also in fewer stoppages, thus reducing noise and pollution.

Parking restrictions have tidied up the effect of indiscriminate use of the streets by stationary vehicles, but in some circumstances the improvement in visual amenity has been offset by the proliferation of signs and markings, parking meters and barriers.

Traffic management has played a big part in reducing congestion, and it is largely due to such measures that journey times of traffic in cities have been reduced, in spite of large increases in the volume of traffic. Perhaps, the very success of these measures has delayed the formulation and execution of radical policies for dealing with city transport.

Restraint of private-car traffic

A possible measure for the reduction of the impact of traffic on the environment in congested centres is to reduce the volume of traffic or exclude traffic altogether on certain streets or areas in a town centre.

There are several ways in which this restraint or exclusion may be applied. The limitation of parking spaces clearly reduces the number of journeys which can be made by a private car with a destination in the area concerned. The street network can be modified by traffic management measures in such a way that journeys by car to a particular destination can be made so circuitous or lengthy in terms of time that drivers are deterred from making such journeys by car.

In some situations, in some cities, access by car to certain streets or areas may only be gained by drivers who posses a licence granted for the purpose. Such systems are at present generally limited to residents who require space at or near their homes, but considerable discussion has centred round proposals for licensing systems for other vehicles entering specified zones and using roads within these zones.

Proposals have also been considered for charging private-car users a sum proportionate to their use of roads in town centres and proportionate to the congestion costs which they would impose on different streets. This pricing of the use of city streets could be achieved by one of two systems.

The first would consist of a meter in the car into which the motorist would insert the amount required in cash or pre-purchased tokens. The meter records the number of pricing points covered. There would be an indication on the car that the meter was in use. Such a system requires surveillance to enforce the use of a meter. The alternative system is one which involves a device in the car enabling the car to be identified uniquely by electronic means when it passes each timing point. A bill is sent to the motorist at intervals.

The basic principle is that the motorist pays for the extra cost he imposes on the community and on other road users, in proportion to his use of city centre streets. Presumably road tax would be reduced so that the total revenue from the use of cars would be distributed in such a way that those who used central areas would pay more than those who did not enter such areas, or did so only rarely. A full and detailed analysis of methods of restraint of traffic in cities may be found in *Better Use of Town Roads* (HMSO 1967).

At present, there is doubt that these automatic devices can be manufactured to the necessary degree of reliability and, in the case of the first method, made sufficiently secure. There may be objections from the public to the second method, because of the ability of the system to exercise surveillance on a private motorist's movements. There is likely to be more general criticism of the basic principle in that the less-affluent car user is affected more than the better-off one. The car user whose expenses are borne entirely by a firm is unlikely to be deterred. Indeed, the latter is less likely to be affected by any monetary policy of restraint.

Town planning measures

The measures so far discussed are for traffic after it has been generated by a fixed disposition of homes, shops, schools, factories, commercial premises, administrative buildings and so on. When planning new areas or re-planning existing areas, it is possible to consider the relative location of different types of land use in relation to accessibility. Such considerations should not be confined to accessibility by the road system only, nor should the road system be thought of in terms of the private car only. Transport and accessibility are far from being the only factors involved. Suitable available land, the requirements for water supply and sewerage systems are all part of a complex process. It becomes a serious issue of policy to consider how far freedom in the location of homes, schools, factories and so on, should be limited in favour of a particular system of movement of people and goods. This issue, together with the implications of other measures discussed above, will be the subject of the next section.

Objectives and policies

The wider context

Transport movements, especially those involving road vehicles, not only have an adverse effect on the built environment but, by reason of their number and complexity in cities, they create a waste of resources through inefficiency arising from the congestion which they themselves create. If the objective of any policy is limited only to the mitigation of these bad effects in purely transport terms, then either the objective will not be fully attained, or other problems will be created in activities other than transport, e.g. if accessibility to a town centre is reduced either by increased congestion through lack of a policy or by a policy directed to that end, there will be a reduction in activity in the centre.

The activities which decline in the city centre will be transferred to more accessible places on the outskirts, a process which has been seen in the extreme in some cities in North America. Can this country, in terms of the limited land available, afford large-scale encroachment on the countryside? Can we afford to allow our city centres to decline till they become urban deserts taken over by the depressed and sometimes the violent? It has already been shown that in some cities a high proportion of congestion in the centre is caused by vehicles whose movements are unrelated to the centre's activities, but whose journeys would be drastically affected by measures exclusively directed to stopping traffic entering the centre.

Considerations such as these led to the opening statement that transport objectives and policies cannot be considered in isolation from the wider context of the city's form and activities. The policy-maker should begin by examining the objectives to be sought for the city as a whole, as these are the prime determinants of the amount and nature of transport movements.

Policies in isolation

Many people see the problem of city congestion as created by the commuter who travels to the city centre by car to work. We have already seen that such journeys do not constitute all of peak-hour travel and that there remains a considerable amount of travel outside peak hours. Several policies have been tried or proposed to reduce the use of the private car in cities, the implicit assumption being that if people are prevented from driving to work in the city centre they will transfer to public transport. For efficiency, bus and rail transport must have individual journeys consolidated into vehicle loads or, if not full-vehicle loads, a sufficient load factor must be achieved to make the service pay. It is a characteristic of most cities in the UK that a large proportion of the population of all social classes lives in suburbs with relatively low density of building. Such suburbs, being peripheral, have large areas remote from major arterial roads, and it is difficult for passenger journeys to be consolidated at the suburban end of the journey. It is only at the city end that consolidation is achieved. In the case of cross-town journeys, which are much more diffuse in pattern, there is difficulty in consolidating journeys at each end. It is this diffusion of pattern which makes it difficult, if not impossible, for public transport to serve with any degree of efficiency. It becomes, therefore, a matter for consideration whether or not, from a community point of view it is more efficient to have such journeys made by private car. The construction of new roads which avoid the centre would deal with such traffic.

A major problem is peak-hour travel. If private-car travellers are to be prevented or discouraged from using their vehicles for the journey to work, and instead travel by bus or train, an already serious problem for public transport operators will be accentuated. As things are at present, a major source of loss to bus operators arises from the need to maintain a fleet of buses and their crews to meet peak demand, while for the rest of the day they are running with few passengers, or indeed in some cases are idle for several hours in the shift. Any major transfer of passengers from cars to buses at peak hours will make this unprofitable position even more serious.

One possible way in which the difficulties can be eased without public capital expenditure is by the extension of the use of flexible working arrangements ("Flexitime"), or even arrangements for staggered hours between adjacent establishments. In one town where staggered hours were introduced at industrial estates, peak traffic was reduced by 12% and as some buses could make more than one run, the bus operator was able to serve the establishment with fewer buses. A reduction of 12% to peak-hour traffic results in an increase in speed of traffic in an average city street of 2 miles/h; this is substantial in streets operating at present at speeds of 12–15 miles/h on average. Flexible working arrangements on a widespread scale would be even more beneficial than staggered hours. Although no public capital expenditure is involved, some expenditure is necessary on the part of the employers in connection with the installation of time recording systems and pay computations systems. This has been shown to be offset by increased productivity and reduced turn-over of staff.

The use of "park and ride" systems to get over the difficulty of low-density travel in the suburbs is a subject of mixed reports. It would appear that, if time spent in changing the mode of transport is high in relation to the overall travel time, there is little inducement to use such systems. This is not just a reaction on the part of private-car owners, but has been observed in systems where buses connect with trains using one through ticket when the overall journey is relatively short.

Travel by bus is still by far the commonest form of travel to city centres, and clearly everything possible must be done to improve the efficiency of such services for the benefit of existing travellers. Moreover, the aged, the infirm, those who are too young to be allowed to drive, and those who do not wish to own a car, even when able to do so, must be catered for. It remains, therefore, the case that a major objective of any transport policy must be the maintenance and improvement of bus services and, where applicable, rail services in our cities.

However, we must not be under any illusions as to how far any major transfer of car passengers to buses can be achieved or, indeed, is desirable. As long as people aim to live at low densities, public transport will be at a serious disadvantage.

So far as measures of new road construction are concerned, it has already been pointed out that such new roads help bus services, whether they operate on the new roads or not. But the construction of new roads cannot be planned in isolation, because of the effect of greater accessibility on mobility?

Do we need greater mobility?

Modern society requires a high degree of mobility to function efficiently. Therefore, reduction in mobility can only be achieved by some change in our way of life. Reasons for the increasing demand for mobility are the demand for more space for living, and the growing concentration of industry and shopping in larger units—the so-called *economy of scale* principle.

Even where industry is small-scale, modern planning concentrates all the small-scale industries on one site—the industrial estate. This may result in some saving in services (though this is questionable), but a greater dispersion of smaller industries might well reduce the amount of travel. This policy requires re-thinking, but any solution must balance total costs, benefits and resources, irrespective of whether these accrue to the community or to private interests.

The benefits of economies of scale are being increasingly questioned.

From a transportation point of view, the concentration of activities in a few intensely used locations, taken together with the increasing dispersal of housing at low densities, leads to greater pressure in the use of the private car, and to increased need for large-scale facilities near the points of concentration.

It has been demonstrated by Sir Colin Buchanan, Professor R. J. Smeed and others that mobility by private car decreases with the size of the town. There may well be good social and economic reasons for limiting the expansion of cities, and overall planning policies in this country have been directed to this aim in conjuction with the establishment of New Towns. A reduction in the intensity of the consumer-orientated society, with its concomitant large-scale use of resources and concentration of activities in large-scale production, together with a policy of limiting the size of cities, may result in a reduction of the required mobility.

From the individual citizen's point of view, a reduction of the time wasted in travel would be welcome. (Here, travel for pleasure is excluded; although such travel has a serious effect on some areas, it does not as a rule affect the built environment.)

What kind of city?

Clearly, all consideration of policies for transportation must take account of the complexities of interaction which have as their basis the analyses given in Tables 3.1 and 3.2 and figures 3.1 and 3.2.

The objectives of transport policies must be consistent with the overall planning and economic objectives for city life. The transport problem cannot be attacked fundamentally, nor can transport planning for the future be successful, unless the nature of the future is defined.

Who is to specify the nature of the city of the future? Architects and planners for some generations have described the "ideal" city and the conditions necessary for happy and harmonious community living. However, there seems to be no agreement in terms of the physical expression of the "ideal" city. Even with a greater measure of agreement among trained and skilled professionals, what evidence is there that the public would be in agreement? Consider, for example, the extreme change in housing policy from single-level housing to multi-storey living, and now the revulsion from the latter style. There appears to be no way of knowing the kind of city wanted by the public, still less of knowing whether the public are prepared to pay the price for achieving their ideal. Besides, any ideal includes imponderable qualities which cannot be measured in economic terms.

It is doubtful if the public appreciates the powerful forces of change which are at work in cities—due in a large measure, though not exclusively, to modern communications; inherent resistance to change encourages people to accept the state to which they have been accustomed. Major physical changes are resented and attributed to the "planners"—an omnibus term by no means reserved for professional town planners.

Although there is the difficulty of determining the kind of city that people want, some tendencies seem to emerge. People seek to satisfy two conflicting requirements: a home in near-country surroundings, and the facilities of a large busy city centre. This is the inherent problem of public transport.

If we continue to build low-density suburbs, then economically self-supporting mass public-transport systems are impossible. There arises the question of whether we should build in higher densities so that we can base movements in cities on public transport. While there are social problems involved in high-density building, low-density areas also have problems, such as isolation for wives at home.

The major issue of our time has been most succinctly expressed by Peter Hall in *Developing Patterns of Urbanization*:

> As city regions disperse on the regional scale, are they to reconcentrate on a local scale? How far should localized high-density modes of activity, linked to high-density public transport but with inevitable congestion limiting private transport, be permitted and even encouraged in the growing outer peripheries of our great metropolitan regions? This, above all,is the central question for urban organization in Britain, and in Europe, for the years up to 2000.

The fuel crisis

How far might all predictions and policies on transportation be affected by the fuel crisis? Experience shows that, for car travel, the pressure to have the advantage of personal mobility is so powerful that users will adjust to the higher price as they have to the higher prices of other commodities. The improvement and strengthening of public transport may well be assisted by a greater use of its services by car owners—but even public transport costs will be affected. The exhaustion of oil resources for transport is a minor problem compared to its effect on industrial production. Industry may revert to coal-based production, but equally transport may do the same. Some very efficient steam road vehicles were developed in the nineteenth century, but progress was soon stifled by discriminatory taxation.

If the standard of living is to be maintained by the use of other forms of energy, then the need for mobility will remain, and technological ingenuity will find some form of alternative fuel. If this is too optimistic a view, then the alternative will be that we will not have the mobility necessary to maintain the kind of industrial production which gives us our present standard of living. Meanwhile, we have to live in our cities, and measures to make them more habitable are still worth while for the sake of this and several more generations.

Acknowledgments

A list of selected works for further reading is appended to this chapter, but the number of sources consulted is very much greater than is represented by the list. Moreover, the author's view have been shaped and modified by the works of others over the years, and it is likely that some of the influences derived from these sources could not be indentified at this stage. To all the authors and research workers in this complicated and controversial field of urban transport, whose work has contributed, wittingly and unwittingly, to this chapter, the author offers his grateful thanks. He also wishes to thank the many people who, by their questions and criticisms, have stimulated him into re-thinking his views on the subject of urban transport.

FURTHER READING

1. *The City in History*, Lewis Mumford, Secker and Warburg 1961, Pelican Books 1966.
   Mumford's purpose, to quote his own words, is "to deal with the forms and functions of the city, and with the purposes that have emerged from it". The fulfilment of these purposes gives rise to transport which has its impact on the city. This work is a most valuable foundation study of the origins and nature of the city, and provides an understanding of the context within which the transport problem should be studied.
2 *Developing Patterns of Urbanization*, (Ed.) Peter Cowan, Oliver and Boyd 1970.
   This book consists of a collection of papers by various authors in which the results of the Centre for Environmental Studies Working Group were published. Planning involves forecasting, and as a preliminary to any attempt to forecast the future, existing trends and pressures for change must be identified. Although the paper by Peter Hall on "Transportation" is the most directly relevant to the subject of transport and the built environment, the other papers are of great importance in demonstrating the changes at work in urban society. These changes will affect the quality of urban life, and in turn the kind of movements which people will make.
3. *The Exploding City*, (Ed.) W. D. C. Wright and D. H. Stewart, Edinburgh University Press 1972.
   It contains a record of papers and discussions at a seminar on Urban Growth and the Social Sciences held at the University of Edinburgh. Although apparently covering the same ground as "Developing Patterns of Urbanization" there is sufficient difference in approach to make this a complementary work. The inclusion of discussions on the papers serves to stimulate the reader and emphasizes unresolved issues which the community has to face.
4. *Planning for Man and Motor*, Paul Ritter, Pergamon Press, 1964.
   Beginning with the man-vehicle relationship, Ritter examines the needs of both, relating them to psychological and sociological aspects. The relation to the environment underlies the whole treatment. Plans (as they existed when Ritter was writing) for New Towns, urban renewal and traffic segregation are illustrated and analysed, examples being given from all over the world. The book is vigorous and well illustrated and, although some ideas and techniques have changed since it was written, the fundamental concepts remain. This is an excellent and most comprehensive study for the layman.

5. *Urban Survival and Traffic*, (Ed.) T. E. H. Williams, E, and F. N. Spon 1962.

The symposium, the proceedings of which constitute this volume, was held at the University of Newcastle upon Tyne in 1961 and was a landmark in the evolution of the professional approach to the problem of traffic in cities. Traffic and highway problems had been the subject of many conferences, but the "Newcastle Symposium", as it came to be known, brought together representatives of the three professions of town planning, architecture and civil engineering from all over the world, and recognized the interdependence of the three professions. Although some years have passed and many of the aims discussed at the symposium have been achieved and techniques have advanced, there remains much to interest the general and specialist reader alike.

6. *Traffic in Towns. A Study of the Long-term Problems of Traffic in Urban Areas*, HMSO 1963.

The Reports of the Steering Group and the Working Group appointed by the Minister of Transport form this Report, popularly known as "The Buchanan Report". As a statement on the impact of road traffic on the urban environment this Report remains unsurpassed. There have been professional criticisms (made with hindsight conferred by developing techniques) of the processes used for forecasting, but the basic approach proposed remains unchallenged. This approach suggested that the kind of environment which the community wants in a particular area must be defined. Once the definition is made, all extraneous traffic must be isolated from the area. The Report also made an effort to measure environmental effects in quantitative and economic terms, a process which has been extended since. Whether the reader agrees with all the views in the Report or not, it remains fundamental reading.

7. *Roads*, Alan Day, Mayflower Books 1963.

This is a stimulating, lively and provocative discussion of the problems of traffic in towns. Most of the issues raised by Alan Day are as relevant today as when the book was published. All of the analyses and contentions are supported by facts and figures, but these details are never permitted to interrupt the flow of reasoned argument. Excellent reading for the non-specialist.

8. *New Roads in Towns*, Report of the Urban Motorways Committee, HMSO, 1972.

The Urban Motorways Committee identified the adverse effects that new urban roads might have on people living close to them. The Committee's Report co-ordinates and extends much of the work which has already been done on the subject but, in particular, added to architectural and engineering solutions proposals for administrative reforms which would remove many of the constraints previously imposed on engineers who were conscious of the problems on the impact of highways on the environment but lacked the administrative powers to deal with these problems by means other than purely engineering measures. Emphasis is also given in the need for the joint inter-professional approach to location and design as was first highlighted by the Newcastle Symposium (Reference 5 above).

9. *Better Use of Town Roads*, HMSO 1967.

It is unlikely that, in larger towns and cities, new construction alone can relieve problems of congestion and associated environmental effects. Control of traffic on the remaining network will be necessary, and for many cities restraint on the use of the private car may be required. This report is a comprehensive and thorough examination of possible means of restraint and is very realistic in its consideration of means of implementation and enforcement of policies of restraint.

10. *Effects of Traffic and Roads on the Environment in Urban Areas*, OECD, 1973 (sales agents in UK, HMSO).

This is a valuable summary of the problems created by noise, pollution and visual intrusion. The scope of the Report is not limited to the United Kingdom, but comparison is made between many countries facing the same problems. Although the Group which produced the Report is a Research Group, there are no technicalities involved which cannot be understood by the layman. The terse writing has enabled a large amount of valuable and basic information to be packed into a mere 60 pages or so and there are over 40 references to more detailed studies.

CHAPTER FOUR

SOLID AND LIQUID WASTES

JAMES McL. FRASER

Introduction

Wastes . . . to the archaeologist, a revelation; to the clinician, a vector in disease; to the ecologist, a disruption; to the aesthete, an offence; to the conservationalist, a wasted resource; to the producer, a harassment; to the processor, a source of profit. Wastes are an inescapable concomitant of living. They originate in the body, the home, the factory and the field.* Their nature and characteristics are as varied as the sources whence they come.

General description

The influence which wastes exert on the environment, and the means which are adopted for their management, depend more on whether the wastes are solid or liquid, and less on where they originated. This convenient, if not altogether rational, classification is followed in this chapter.

Liquid wastes

The one liquid or near-liquid waste that is produced throughout the world in more or less uniform quality and quantity is human excreta. Each adult produces around 150 g of faeces and 1500 ml of urine per day. Although the

* Radioactive wastes and wastes in relation to atmospheric pollution do not come within the scope of this chapter.

faeces contain some undigested food, their composition and characteristics are not as affected by variations in diet as might be thought likely, the explanation being that a large part of the faecal mass is not of dietary origin. Living and dead organisms may account for about one third of its total weight.

Also produced throughout the world are the liquid wastes from such domestic activities as personal hygiene, the preparation of food, the washing of clothes, and the general cleaning that is a part of any domestic scene. By contrast to the near-uniformity of human excreta, these domestic wastes vary widely in quality and quantity from country to country. The most important factors are the availability of water and the prodigality with which it is used, but social, religious and ethnic practices also exert their influences. In the countries in which the usage of water is highest, the domestic wastes of the kind just described may amount to 600 litres per person-day.

Another form of liquid waste is stormwater. This can be defined as the fraction of the total rainfall that falls on impervious surfaces such as roads, roofs, car parks, school playgrounds and the like. Older parts of a town centre may be almost wholly impervious, and over a large area of mixed urban development almost half of the total area may be impervious. There are two problems associated with stormwater: its polluting properties and its volume. Although rain that has passed through an urban atmosphere is liable to be polluted by carbon dioxide and sulphur dioxide, it is, in relative terms, clean when it reaches the ground. There it quickly becomes polluted with oil, grease, rubber, lead from leaded petrol, grit from chimney emissions, chemicals which have been spilt on factory storage areas and, if there are animals around, their excreta. The second problem associated with stormwater is one of intensity. If the present pattern continues, it is to be expected that at some time in the future there will be a rainstorm so severe in its intensity that it exceeds all rainstorms that have preceded it. Because of these uncertainties about both its quality and quantity, stormwater is a waste material which presents unusual difficulties.

Another source of liquid wastes which affects the built environment is industry, a term that is capable of being interpreted differently from country to country. In the United Kingdom its statutory definition includes, in addition to the processes of manufacture, such activities as agriculture, horticulture, scientific research or experiment and the carrying on of a hospital or nursing home. As any industrial waste is liable to contain raw materials, catalysts, solvents, intermediates, final product and detergents, there is virtually no limit to the variety of characteristics which industrial wastes display. Some are acid, others alkaline; some toxic, others not; some polluting and pathogenic, others merely polluting; some predominantly organic, others inorganic; some containing solid matter in suspension, others with solid matter in solution, still others containing no solid matter at all; some crystal-clear, others coloured or turbid; some (for example, fermentation wastes) characteristically consistent in quality, others (from fruit and vegetable processing) markedly inconsistent.

Even if it were practicable to do so, it would be unavailing to describe any liquid industrial waste in detail. Nevertheless the leading characteristics of a few wastes are as stated in Table 4.1.

Table 4.1 Typical characteristics of liquid wastes from selected industries

Malt whisky production
Large quantity solid matter in suspension; acidic; strongly polluting; complete absence of pathogens; various metals in solution.

Dyeing
Acetic or sulphuric acidity; strongly coloured; temperature almost 100°C; no solid matter in suspension; complete absence of pathogens.

Cholera hospital
Large quantity solid matter in suspension; pathogenic bacteria of several species of the order of thousands per ml; *Ascaris* ova of the order of thousands per 100 ml; acidity normal; highly offensive.

Tannery
Part caustic: part acid; highly polluting; large quantity of solid matter in suspension; much of it putrescible.

Piggery
Highly polluting; large quantity faecal and fibrous matter in suspension; pathogenic; large quantity copper; very odorous.

Although it may seem inappropriate to refer to agricultural wastes in a chapter whose main theme is the built environment, the incidence of agricultural wastes on the built environment is becoming more significant as methods of animal husbandry become more intensive. Two or three acres of land are sufficient for 1000 pigs and one acre of factory floor is sufficient for 84 000 head of poultry. The difficulties of disposing of the dung and droppings from such densities of animal populations can be immense, and it may be the capacity of the waste disposal facilities and not considerations of animal husbandry which decides how many animals will be kept. Consider, for example, the case of 1400 cows confined on three and a half acres of land. If their dung were spread at the normal rate of application, about 3000 acres of land would have to be made available for its reception. Apart from the unlikelihood of sufficient land being available, the cost of storing, transporting and spreading such a low-value commodity might well be the deciding factor in the economic viability of the scheme. An agricultural waste of a different kind is the liquid which oozes out of silage if the green crop is placed in the pit or tower directly after it has been cut. This liquid is highly polluting, but its quantity can be reduced and its characteristics modified by the crop being allowed to lie in the field until it has wilted.

Solid wastes

In general the types of solid wastes that are produced in industry have little similarity to those produced in the home, but there are two noteworthy exceptions; packaging materials, and discarded vehicles and their appurtenances.

Packaging materials consist of drums, cans, cartons, collapsible tubes, squeezy bottles, shrink-wraps and the recently introduced one-trip glass bottles. The problems associated with the disposal of these wastes have been responsible for the introduction of the "over-packaging" concept, defined as the situation which arises when the community as a whole would be better served if fewer economic resources were used in the packaging of a commodity.

The disposal, in the course of a vehicle's useful life, of waste oil, discarded tyres, batteries and other components, and in the fullness of time the disposal of the vehicle itself, is a problem which affects the countryside even more than the built-up areas. If anything can be said in defence of discarded vehicles it is that they are neither pathogenic nor polluting in the restricted sense of the term. The suggestion is attributed to a President of the United States that in the purchase price of every vehicle there should be an element that would be used for its ultimate destruction.

The wastes that arise from within the house—called refuse or garbage—vary in their characteristics from country to country. In the United Kingdom refuse reaching the point of final disposal is quite likely to contain colour television sets, bicycles and clothing that has been discarded for no better reason than that it has become unfashionable. By contrast, domestic refuse from some Far East communities consists of little more than a soggy mess of fruit and vegetable skins and parings. Every piece of waste material that can be sold or reused or converted to some other use is painstakingly searched for and removed. Waste paper, which accounts for about half the weight of refuse in the United Kingdom, is removed, every scrap of it, from the refuse in some developing countries. For reasons such as these, there is nothing approaching the uniformity of quantity and quality that was mentioned in relation to human excreta.

A characteristic of domestic wastes in the United Kingdom is the manner in which their composition is affected by fashion. The incidence of discarded clothing has already been referred to, but fashion also affects other constituents of domestic refuse. Beverage containers, for example, are chosen from among glass bottles, plastic bottles, "tins" or aluminium cans largely according to the purchaser's preference. Even shoppers' carrier bags are affected. At one time they were paper, latterly plastics; now, about to be introduced, is the carrier bag *de luxe*. It is of biodegradable polyethylene which, in contact with bacteria or fungi in the presence of moisture, will slowly disintegrate. This deliberate shortening of the life of the material has been brought about by incorporating grains of starch in the plastics. It is the starch which is attacked by the bacteria and fungi, not the plastics material itself.

If season and social distinctions are disregarded, the composition of domestic refuse in the United Kingdom is as stated in Table 4.2.

Table 4.2 Typical composition of domestic refuse in the United Kingdom

	percentage by weight (round figures)
Metal	6
Glass	6
Paper	50
Rag	3
Vegetable and putrescible	13
Plastics	2
Dust and cinder	18
Unclassified	2
	100

Refuse with the foregoing characteristics has a bulk density around 115 kg/m^3 (8 lb/ft^3) and its daily "arisings" (i.e. production) is approaching 1 kg (2·2 lb) per person. Such a refuse is autothermic, i.e. capable of sustaining combustion without the addition of heat from an external source. It has a gross calorific value of around 9000 kJ/kg (4000 Btu/lb) but there is some evidence to suggest that the calorific values of refuse here and in USA are tending to fall.

For comparison, the calorific values of some other combustible materials are given in Table 4.3.

Table 4.3 Calorific values of some combustible materials

	kJ/kg	Btu/lb
Coal	32 000	14 000
Wood	16 000	7 000
Peat	16 000	7 000
Paraffin (kerosene)	42 000	18 000

Because of the great quantities in which they are produced and their prominence in the landscape, non-domestic solid wastes might be thought of as being chiefly the wastes of the extractive industries. Examples would be the quartz and micaceous wastes from china clay washings in Cornwall, the slatey wastes from quarries in Wales, Cumbria and the West of Scotland, and the pithead wastes throughout the coalfields. There are, however, other wastes which are environmentally significant, among them being the red blaes from the distillation of oil shale in central Scotland, the red mud from the processing of bauxite which is the mineral from which aluminium is derived, the calcium sulphate produced in the cause of phosphoric acid manufacture, and metallurgical slags of various kinds. The waste in this category which has probably the widest distribution is pulverized fuel ash (PFA) which is the residue from the burning of pulverized solid fuel in electricity generating stations and elsewhere.

There are, however, solid wastes which, in quantity, are insignificant by comparison with either house refuse or industrial wastes, but whose effect is not without significance. In an intensive egg production unit, the bodies of 200 dead hens may have to be disposed of each day. So must sawdust and shavings from woodworking; trimmings of plastic upholstery for the automotive industry; feathers, hooves, hair and condemned meat from the abattoir, and many others. Not all domestic wastes find their way to the refuse sack. Special disposal arrangements have to be made for placentas from babies born at home and for the wastes from domiciliary renal dialysis machines. Likewise some of the wastes that do find their way into the refuse sack may give rise to unexpected difficulties, colostomy bags, babies' disposable nappies and incontinence pads being examples.

Effects of wastes on the built environment

It cannot be stated categorically what injury is caused to the built environment by the presence or production of wastes, because only some of the effects are tangible and capable of being quantified, whereas many are intangible or produce a wholly subjective response. Nevertheless, it is not inappropriate to consider the effects of wastes under two heads: as hazards to health, and as causing pollution and loss of amenity.

Hazards to health

If, as the World Health Organization define it, health is "... the state of complete physical, mental and social wellbeing ...", consideration of the hazards to health would be incomplete if it did not take into account the resentment and repugnance which result from the impingement of someone else's wastes on one's own peace and privacy. Relevant as this may be, it is too far removed from the main theme of this chapter to allow it to be pursued. For the present, the hazards to health which are being discussed are chemical or biological hazards, not psychological ones.

A distinction must be made among wastes that are polluting, pathogenic or toxic, or that exhibit more than one of these characteristics. Human excreta are potentially pathogenic, in addition to being polluting. The wastes from the manufacture of whisky are polluting, but otherwise clean. Wastes from metallurgical processes may be toxic by reason of the cyanide or arsenic which they contain, but they are biologically inactive. In any examination of the effects of wastes on human health, none of these differences can afford to be overlooked.

Although the ingestion of chemically polluted water or food has resulted in much suffering and loss of life, these occurrences have been of relatively local significance. The poisonings which befell a Japanese fishing community in the 1950s and 1960 are a classical example. The deaths and sicknesses were attributable to the eating of fish and shellfish which has become polluted with dimethyl mercury $(CH_3)_2$ Hg, a waste from the mercuric catalyst used in the manufacture of acetaldehyde. This particular mercurial compound is capable of being accumulated in the flesh of fish and shellfish to a level that is some hundreds of times greater than its concentration in the sea. Other injurious substances can also be accumulated in the same way, arsenic and cadmium being noteworthy examples.

A much more widespread health hazard results from the ingestion of pathologically active material, the chief source of which is human excreta. Diseases which may be communicated by this means are:

Bacterial diseases: cholera, typhoid, dysentery.
Viral diseases: poliomyelitis, infectious hepatitis.
Parasitic diseases: amoebic dysentery, worm infestation.

The secretions from even a healthy individual are potentially injurious to another but, if the excreta come from one who is himself a sufferer, widespread havoc can be wrought. When it is realized that the number of bacteria alone, i.e. not including viruses or parasites, which are excreted by a person in normal health may amount to several million per gram of faecal matter, the wonder is that anyone can escape. The simple explanation is that many of the bacteria have difficulty in remaining alive out of their accustomed environment. This unfortunately is not true of the viruses or the parasites, some of which are extraordinarily resistant and capable of surviving for years in apparently uncongenial surroundings.

Cholera is a classical example of a disease whose principal vector is sewage. The disease is endemic in only a few areas in the world, Bangladesh being possibly the most seriously affected of them all. It is not difficult to understand why this should be, for the most elementary personal hygiene is not practised in a way that is conducive to disease-free living. Even in the cities, there is an inherent abhorrence in the minds of some of the people to excreting in a place set aside for that purpose, especially if it is at all odorous. Moreover, especially in the rural areas, wholesome water is not available, even for the most restricted dietetic purposes, and the use of heavily polluted water for cooking and for washing the utensils used in cooking is commonplace. Cholera is not, however, unknown in highly developed countries. In Italy within recent years an outbreak of cholera claimed several deaths, the cause being believed to be the consumption of fish which, by living in sewage-polluted sea water, had accumulated in their bodies sufficient cholera *Vibrio* to infect the eaters of the fish.

It is now realized that the long-established custom, in China and other eastern countries, of feeding human excreta to swine exposes the persons who consume the pork to an unjustifiable health risk; so, too, does the use of untreated sewage to irrigate land on which crops are grown or animals are fed. The risks associated with the use of sewage that has undergone effective treatment are much less, but even so its use on fruit or vegetables which might be eaten without being cooked is undesirable. Examples of such crops are soft fruits, lettuces, cucumbers, tomatoes and the smaller salad crops.

One risk to health from exposure to human wastes has for long been presumed to arise from bathing and participating in aquatic sport in sewage-polluted sea water. The presumption that such a risk existed had not been supported by adequate bacteriological or epidemiological evidence until the Public Health Laboratory Service, after a five-year study of material from England and Wales, reported in 1959 to the following effect. Although faecal coliforms and *Salmonellae* serotypes were isolated from sea water in substantial numbers, the conclusions were that even beaches that were aesthetically very unsatisfactory carried only a negligible risk to health, and that the risk would not become serious unless the water was so foul as to be aesthetically revolting. These conclusions did not commend themselves to everyone, some holding the view that mental or physical revulsion consequent upon contact with recognizable faecal matter was as much a health risk as a dose of diarrhoea. In the United Kingdom, there is no yardstick by means of which an area of the sea may be adjudged suitable or unsuitable for aquatic sport, but the European Economic Community (EEC) have recently issued a draft directive that no fresh water or sea water in which bathing is permitted shall fall below standards based on microbiological, physical and chemical characteristics.

Some countries have already prescribed such standards, but there is no international agreement. According to one standard 10 000 coliforms and 200 *Streptococcus faecalis* per litre of sea water would not be unacceptable.

There is no evidence in this and other developed countries that persons who are exposed to contact with domestic refuse are, health-wise, at a disadvantage in relation to other members of the community. In some developing countries, however, the incidence of intestinal parasites is much higher among refuse workers than among other groups. The inference is that the risks to which all refuse workers are exposed are similar, but that a high standard of personal hygiene is sufficient protection.

The risks associated with industrial solid wastes vary from the acute toxicity of cyanide to the insidious and possibly carcinogenic effects of such substances as lead, silica and asbestos.

Pollution and loss of amenity

Excreta are complex chemical substances which consist mainly of carbohydrates, proteins, fats, amino acids, fatty acids and water. The effect that they and other predominantly organic wastes have on the environment arises from their predisposition to take in oxygen. If, for example, such wastes are discharged into a body of water, the water is deprived of its oxygen and, depending on the circumstances, the oxygen depletion may approach the stage at which plant and animal life cannot survive. The ability of a waste to abstract oxygen in this way is called its biochemical oxygen demand (BOD). BOD is central to the thinking on quantifying pollution, comparing the polluting capabilities of different wastes, and assessing the probable consequences of discharging a waste into a body of water.

Although it has been stated that the characteristics of human excreta are more or less uniform throughout the world, there are, as might be expected, differences between the excreta of meat eaters and of vegetarians, and between the excreta of those who enjoy a high standard of living and those who are on the verge of starvation. In most western countries the BOD attributable to one person's daily wastes is around 54 g (0·12 lb). The effect of discharging one person's wastes into fresh water may be deduced from the fact that at 10°C, fresh water that is fully oxygenated contains 11.3 mg/l* of oxygen. Hence the wastes that one person produces in one day can remove all the oxygen from 4750 litres (1045 gallons) of oxygen-saturated water.

* The smallness of one milligram per litre may be more readily realized when it is translated into more familiar dimensions. One milligram per litre is *numerically* the same as one sixteenth of an inch in a mile, or one drop in 55 gallons, or thirty-one seconds in a year.

While de-oxygenation is taking place, the water makes an effort to re-oxygenate itself by absorbing oxygen from the atmosphere. Whether the eventual result is a lowering of the oxygen level or a return to the initial level depends on many factors, important among them being the relative quantities of the wastes and the water, the oxygen demand exerted by the wastes, whether the water is flowing or stationary, and whether it is turbulent or still. Characteristically, if an oxygen-demanding waste is discharged into a river, the level of the oxygen in the water at the point of entry is not immediately affected. As the water moves downstream, the oxygen becomes depleted but the level starts to rise again as a result of natural re-oxygenation. This phenomenon is known as the *oxygen sag* and is of importance in river management. The discharge of warm but otherwise uncontaminated water into a river brings about a lowering of the oxygen level in a manner similar to that just described.

Oxygen deficiency is much more than a useful indicator for river management. The lower the level of oxygen in the water, the lower the form of life the water is able to support. The first of the fauna to be affected are the sensitive species of migratory fish, such as trout or salmon, next the less sensitive, such as the so-called coarse fish, than the plants which require oxygen in their metabolism, until finally the only living matter is the familiar stringy fungus-like growths commonly referred to as sewage fungus but botanically of the *Sphaerotilus* genus. With it there may be some *Tubifex* worms which can exert an extraordinary resistance to pollution.

De-oxygenation, although the commonest, is by no means the only manifestation of pollution, because any happening which upsets the natural regime in the water is liable to bring about its degradation. The discharge of particles of solid matter, possibly innocuous in themselves, can effectively smother the food on which fish depend for their existence and so bring about their extinction. A similar result can be expected from the discharge of wastes that are too acidic or too alkaline to allow natural growth and reproduction to take place, even although the acidity or alkalinity may not be harmful of itself.

Pollution can be especially serious in the passage of mature fish from the sea to the spawning grounds in the headwaters of a river, and the passage of the young fish from the spawning grounds back to the sea. The effect of a polluting discharge into a river may seem to be of no more than local significance, but it may prove to be an impenetrable barrier to migratory fish.

The commonest means of disposing of solid wastes is to deposit them, with or without some prior treatment, on land. Rain falling on and percolating through these deposits may emerge contaminated with material of which the deposit is composed. This emergent water is given the name *leachate* and it is often extremely polluting and offensive. If the deposit is of domestic refuse with its high content of vegetable matter, waste food, floor sweepings and other organic substances, the leachate is invariably obnoxious. If the deposit is of wastes which are predominantly inorganic, the leachate is liable to be acid, alkaline, toxic, or to have some other characteristic which it has acquired in its passage through the deposit.

A hazard of a different kind associated with the deposition of solid wastes on land is that of combustion arising spontaneously as a result of bacterial activity, or in consequence of some chemical or physical reaction. Deposits of domestic refuse are expecially vulnerable, partly because of the very high proportion of combustible matter they contain and partly because of the large quantity of air that is entrained in such objects as squeezy bottles, plastic containers and cartons. Once domestic refuse has been ignited, it is a matter of great difficulty to extinguish the fire other than by isolating it. Burning is almost invariably accompanied by the emission of smoke, grit and odour—an offence to sight and smell.

It would not be unusually difficult to reckon the loss to the community which the presence of wastes in the environment occasions if account were to be taken of only the tangible effects: the value of the land that is occupied, the damage to water supplies, the reduction in soil fertility, and so on. But these are not the only sources of loss, nor even the most important. Any assessment of total loss would entail putting a value on such imponderables as amenity, health and recreation. Figures so produced would be of little practical use, but they would re-affirm that the value of an asset is not always accurately represented by the figure on the price tag.

Management of wastes

There are forces and agencies in nature that bring about decay and decomposition: steel rusts; biodegradable plastics disintegrate; organic matter is oxidized; stones and rocks are fragmented; carrion is consumed. Is there an application for these natural forces and agencies in the management of the wastes of the built environment? Could the natural forces and agencies be intensified or their effects accelerated?

In manned space travel, algae of the *Chlorella* genus are used to reduce the astronaut's excreta to their useful constituents such as water and nitrogen, the water being re-used as such, and the nitrogen as a filler for the reconstituted air. Ingenious and effective as the technique is, it has its limitations, for even in space travel more wastes are produced than algae are capable of decomposing. Waste food containers and discarded clothing come to mind.

Liquid wastes

In the beginning, liquid wastes consisted almost wholly of excreta. The directions governing their disposal were simple and direct:

> Thou shalt have a place also without the camp,
> whither thou shalt go forth abroad; and thou
> shalt have a paddle upon thy weapon; and it shall
> be, when thou wilt ease thyself abroad, thou
> shalt dig therewith, and shalt turn back and cover
> that which cometh from thee: (Deuteronomy 23–12, 13)

As human settlements became bigger, the burial by the individual of his own excreta had to give way to disposal by the community. For the most part this was accomplished by discharging the wastes into any stream or river that would carry them away from the settlement. As little regard was paid to the suitability of the stream, or the effect that its use might have on settlements downstream of the first, the effects, as can be imagined, were calamitous. As long as the communities were small, the situation was not beyond endurance, however, but the progressive deterioration reached its nadir in the early 1800s. Then, under the influence of the Industrial Revolution, towns became overcrowded, living accommodation indescribably squalid, and the people destitute. In an effort to relieve their penury, the people took to storing human or animal faeces inside their houses and, when sufficient had accumulated, selling it to farmers for application to the land. Today the dunghill has disappeared from the urban environment in the United Kingdom, although its counterpart is still to be seen in some developing countries where "wastes" are too valuable to be wasted. Even in the most highly developed countries, however, the rivers continue to be polluted, although the evidence is that a steady if unspectacular improvement is being achieved.

Such is the state of the art that liquid wastes can be transformed into water of dietetic quality, if there is sufficient justification for making the effort. Apart from manned space travel, to which reference has already been made, sufficient justification is always lacking and, for economic reasons, the current emphasis is on disposal as distinct from treatment. For communities situated within reach of the sea, the means of disposal that is commanding most attention is the dispersion of the untreated wastes in the sea. This is a scientific technique that is not to be confused with an earlier generation of sea-disposal pipes which were deemed to be satisfactory if they reached high water mark. The distance which the pipes must now extend into the sea is calculated by reference to the water movement, salinity and temperature stratification, and other parameters. Such a pipeline is likely to be taken at least 2 km offshore and possibly much farther. A refinement of disposal at sea which is practised in a few countries, notably the Netherlands and the United States, is to bring about the separation (by means which are described later) of the material that can readily be separated from the vastly greater quantity of liquid and substances in solution in the liquid. The separated fraction is then pumped out to sea in its own pipeline, which may be only one tenth of the diameter of the pipeline in which the liquid fraction is carried. The smaller pipeline may extend to two or three times as far into the sea as does the larger.

Of necessity, communities which are not within reach of the sea must contrive to transform their liquid wastes into more or less innocuous substances, the purpose being two-fold: to satisfy all or some of the demand for oxygen which the wastes exert, and to reduce the amount of solid matter which the wastes contain. Disinfection, i.e. the deliberate killing-off of pathogenically active material is not normally a primary purpose of treatment, although in exceptional circumstances it may be. An example might be the wastes from an establishment in which living organisms of a disease were used in the coarse of research. The wastes from such an establishment might be disinfected as a matter of course, as a precaution against the escape of some of the organisms.

Treatment is an intensification of the processes of decay and decomposition which are effected by the natural agencies of sunlight, air and bacteria. A circumstance which militates against high efficiency in the treatment of sewage is the fact that the relatively small quantity of polluting material is diluted by a large quantity of water. From the point of view of treatment, this compares unfavourably with, say, the wastes from a chemical factory, where the volume is relatively small and the concentration correspondingly high.

In general the purification of sewage and other liquid wastes is effected in two stages:

(i) the separation of such particulate solid matter as can be separated from the liquid and its subsequent removal;

(ii) the oxidation, usually by biological means, of the dissolved and colloidal polluting matter remaining in the liquid.

The schematic diagram for the treatment of sewage is shown in figure 4.1.

Where the situation so requires, the treatment processes are preceded by the correction of excessive alkalinity or acidity by the addition of appropriate quantities of acids or alkalis, the cheapest of which are sulphuric acid and lime. The separation of the solid matter from the liquid includes both "sinking" and "floating". It may be assisted by a coagulant

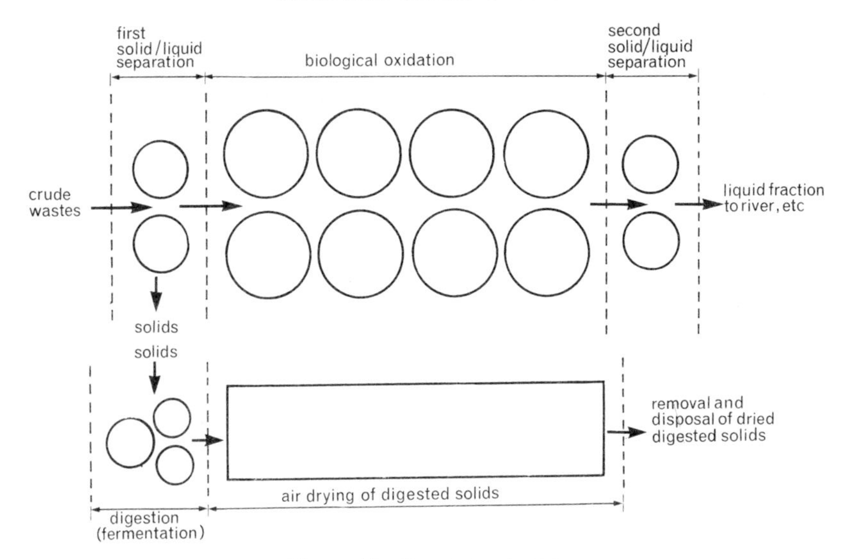

Figure 4.1 Sewage treatment

such as alum, but in the treatment of sewage this would be exceptional, for chemical additives are not customarily employed.

The separation of particulate solid matter may be effected by screens, strainers, centrifuges, filters (which may be pressure- or vacuum-assisted) or flotation (usually air-assisted) but most commonly it takes place gravitationally in tanks whose geometry is decided by the settling characteristics of the solids under consideration. In the treatment of sewage, the tanks are customarily shaped as shown in figure 4.2, although rectangular tanks with a length:width ratio of about 3:1 may also be used. In the case of circular tanks, the sewage enters at the centre and moves radially across the tank towards the outlets, which are situated around the perimeter. The speed of travel to the outlet becomes less as the distance from the centre increases; hence the heaviest of the solid matter sinks to the

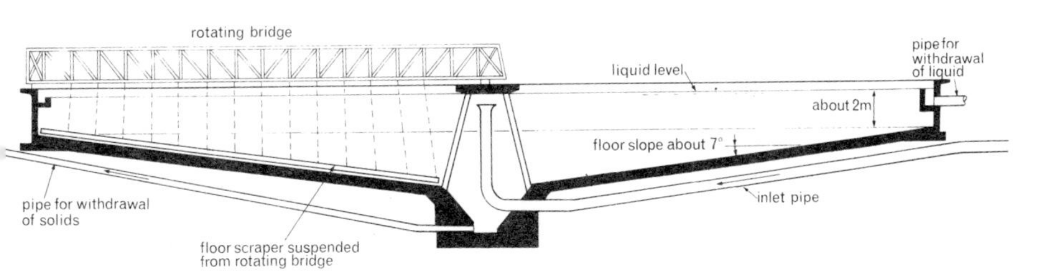

Figure 4.2 Sewage tank

floor near the centre, and the lightest around the outer wall. Slowly rotating scrapers push the solid matter towards the centre of the tank from which it is removed by hydrostatic pressure, suction or other means.

Biological oxidation is effected by bacteria assisted, usually, by some macroorganisms. At the same time as the biological oxidation is taking place, there is some chemical oxidation, but this is small by comparison. The bacteria may be free-moving, in which case they are dispersed throughout the substrate, or they may be fixed to some part of the process structure. In the former event, the number of bacteria constantly becomes depleted by being carried away in the effluent, and this loss is made good by the flocs of bacteria being separated from the effluent and returned to the inlet to the oxidation process, where they mix with the incoming wastes. The oxygen for this form of biological oxidation is customarily supplied by the atmosphere, although the advantages of using "pure", i.e. commercially available, oxygen are discussed later. The air may be diffused through the wastes in fine or coarse bubbles, or turbulence may be induced in the wastes by some form of rotating machinery, in order that the air may be the more readily taken into solution.

In the fixed culture, the bacteria and other organisms take the form of a zoogleal slime. In some cases the process vessel is filled with broken stone, gravel or plastic shapes, to which the slime adheres, and over which the wastes are distributed by a sprinkling device; in others, the film adheres to slowly rotating discs mounted on a horizontal shaft. Only a small sector of the disc is immersed in the wastes; for the remainder of the cycle, the disc and its attached slime are exposed to the air. Whether the fixed or dispersed culture is used, the final operation is the separation and removal from the effluent of the flocs of dead and living bacteria.

Treatment of the kind described is likely to result in the following improvements:

	percentage reduction from crude state
Biochemical oxygen demand	better than 90
Solid matter capable of being separated by settlement	90
Live bacteria	95
Nitrogen	50
Phosphorus	30

If a still higher standard must be satisfied, another stage of treatment is undergone. This may take the form of straining through woven wire of extremely fine mesh, filtration through sand, or irrigation over a grass-covered area of land.

Nitrogen and phosphorus have assumed particular significance in recent years and it is for this reason that they are included in the foregoing table. The presence of nitrogen or phosphorus in a watercourse can set up a state of eutrophication, which is an enrichment resulting from an increase in these elements and other plant nutrients. This enrichment is liable to cause "blooms" of algae and microscopic organisms to develop in such density that light is prevented from penetrating the water, and oxygen from being absorbed by it. The chief source of phosphorus in sewage is domestic detergents, in whose manufacture phosphate is largely used, and the chief source of nitrogen is the oxidized products of urea. Both phosphorus and nitrogen, however, are also wastes of agriculture.

The solids which are separated from the wastes in the first stage of treatment may themselves require further treatment, depending on the nature of the wastes in which they originated. Finely divided coal dust, for example, separated from the wastes from coal preparation, does not require further treatment, whereas the solids produced in the course of treating sewage require to undergo almost as complicated a treatment as does the sewage itself. Sewage solids, collectively called *sludge*, consist of an almost unrecognizable *mélange* of faeces, paper, waste food, grease, soaps and so on. The sludge (which has a most offensive odour) is not highly viscous, but its handling is made the more difficult by its greasiness and lack of homogeneity.

The methods of treating sewage sludge vary from air drying at one end of the scale of sophistication, to digestion (a process which is akin to fermentation) at the other. Between these extremes lie vacuum filtration and pressing (both of which require the addition of a conditioning agent), freeze-drying, incineration and some others. In practice, the choice lies among filtration, pressing, incineration and digestion. Despite its sensitivity to inhibitory chemicals, the last-named is the most common. The end product of the digestion process is a blackish porridge-like substance with a not disagreeable earthy odour. Sludge that has been digested parts with its water readily, so it can be air-dried without difficulty and with little offence. The humus material in the dried sludge can be beneficial to the structure of soils in which there is a humus deficiency, but the presence in the sludge of chemicals, especially those containing copper, lead, cadmium, zinc and selenium, can be very injurious to the fertility of the soil.

Digestion or, as it is usually called in the agricultural context, fermentation can also be used for the treatment of vegetable and animal wastes. Whether it is sewage sludge or vegetable and animal wastes that are being processed, the bacterial conversion takes the following paths:

cellulose, proteins, sugars ⟶ volatile fatty acids
volatile fatty acids ⟶ methane (CH_4), carbon dioxide (CO_2)

Many uninformed assertions are made about the value of digester gas as an energy source. There are many pitfalls and difficulties. The rate at which gas is produced falls off quickly when the temperature drops below the optimum of around 37°C, so in cold weather heat must be applied to the digester to ensure that it continues to function. If no other source of heat is available, the gas itself must be burned, and in this way about one half of the total gas production may not be available for any external use. Occurring as this does at the season of the year when the need for gas is greatest, the situation is somewhat analogous to the eating of the seed corn.

In a well-functioning digester acting on sewage sludge, the gas mixture consists of approximately two thirds methane and one third carbon dioxide. The quantity produced is around 20 litres (0·7 cubic foot) per person-day, and the calorific value of the mixture of gases around 22 300 kJ/m^3 (600 Btu per cubic foot) which is higher than that of coal gas but lower than natural gas. The gas can be burned in the state in which it is produced but, before it can be used in an internal-combustion engine, the carbon dioxide and hydrogen sulphide must be removed. If the gas is to be used in mobile equipment, e.g. in the engine of a tractor, it must first be compressed. In small installations, the cost (not to mention the difficulty) of these operations militates against any use of the gas other than for space or water heating.

Solid wastes

Disposing of the disposables is a formidable task. It is foreseen that domestic solid wastes in the United Kingdom will soom amount to 20 kg per household-week, whereof paper in its various forms will account for about one half. Historically, most wastes have been got rid of (the practice hardly merits any more scientific a description) by being dumped on land or in the sea. A refinement which removes some of the crudity of dumping wastes on land and leaving them exposed to the air goes under the euphemistic term of *sanitary infill.* Its essential feature is that the top surface and the exposed face of the tipped wastes are daily covered by an adequate depth of some inert blanketing material, such as sand, ashes or soil. The cost and difficulty of carrying out this covering up militate against the success of this form of disposal. A recent innovation has been to pulverize the wastes before they are dumped. Pulverization reduces their bulk, hence increases their density—but the difference is one of degree.

The dumping of domestic solid wastes is always accompanied by some risk, irrespective of whether the wastes are dumped crude (i.e. in the state in which they are collected) or whether they have been pulverized. The dump may become ignited, spontaneously or through some external agency;

odours may emanate from it; vermin may be attracted to it; and there are problems associated with leachate which have already been referred to. Nor has dumping at sea much to commend it. Under the influence of tide, current and wind, much of the dumped material may be returned to the land, where it can be identified by its intrusion on the beaches. Even if almost indestructible objects such as plastic squeezy bottles are first pulverized to destroy their buoyance, their shredded remains are not easy to disguise.

Incineration of house refuse, preceded by the hand-picking of rags, glass, metal and paper, was at one time regarded as the ultimate in scientific destruction of the discards of society. Incineration is on the ascendancy, but hand-picking has all but ceased, because of the uncongeniality of the work and the high level of wages that must now be paid. Because of the ease with which ferrous metal objects can be removed from the refuse by magnetic means, this is still commonly practised. The statutory requirements in some countries are that such metal objects must be sterilized before they are baled or removed from the premises. This necessitates passing them through a rotary kiln, and the cost of doing so is casting doubts on the economic advantage of reclaiming even this fraction of the refuse. Glass, even if there were a simple method whereby it could be extracted from the wastes, would present problems of its own, first because of the mixture of colours which the use of reclaimed glass would introduce, and secondly because of the convenience, cleanliness and relative cheapness of using virgin material. Similarly, the recovery of dirty waste paper from among house refuse is of doubtful value to the community.

Objections to the incineration of wastes are almost invariably associated with concern at what may be emitted from the chimney. This may take the form of particulate matter, or odorous or corrosive gases. Odorous gases are evidence of incomplete combustion, and can usually be prevented from reaching the chimney if the furnace is operated at a temperature around 1100°C. The corrosive gas which is most commonly encountered is hydrogen chloride (hydrochloric acid gas) which results from the burning of polyvinyl chloride (PVC). In the United Kingdom plastics seldom amount to more than one or two percent of the weight of domestic refuse, and of the total amount of plastics PVC accounts for a relatively small and diminishing fraction. With such amounts, the problem associated with the release to the atmosphere of HCl resulting from the burning of domestic refuse is not serious. Particulate matter can be arrested by an electrostatic precipitator which, operating at voltages of 20 000–80 000, imparts an electric charge to the particles, which are then attracted to an electrified plate from which they are dislodged mechanically and collected for removal.

The residue of incineration is not as innocuous as might be expected of something that has undergone purification by fire. The total destruction of organic matter is seldom achieved and consequently the residue, although consisting for the most part of ash and inert material, may also contain at least 5 percent of unconsumed organic matter. This is sufficient to make the residue unsuitable for use as a building material. Another constituent of furnace residues is glass. Whether it has been unaltered in its passage through the furnace and emerges as recongnizable splinters and chips, or whether it has become a lump of ashy matter adsorbed on to a fused glassy core, depends on the operating temperature of the furnace. In either event, the presence of the glass makes the residue useless as a cheap infilling material under play areas and sports fields. Most damaging of all, the residue is likely to contain metals, such as lead, copper, zinc and chromium, which could upset the fertility of agricultural land to which the ash might be applied in the belief that it would improve the soil structure.

Interest is being shown in a recently introduced method of disposing of combustible wastes and combustible sludges (chiefly, but not exclusively, sewage sludges) by incinerating them simultaneously. This can be done either in the same furnace or in different furnaces acting in unison. The schematic arrangement of simultaneous incineration in the same furnace is shown in figure 4.3. The sludge, which may have a moisture content as high as 92 percent, enters the top of the furnace and is moved across the hearths under the action of slowly rotating ploughs. It drops from hearth to hearth through openings which are not situated one above the other. The solid

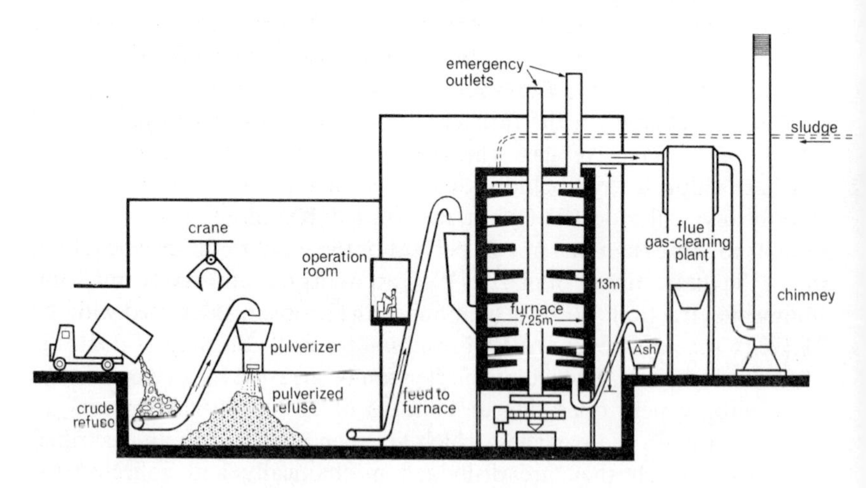

Figure 4.3 Simultaneous incineration of combustible waste and sludge

wastes, which are first pulverized, enter the furnace about mid-height and move across the hearths and from one hearth to another in the same way as the sludge. The hot gases rising from the burning wastes dry the sludge to a state at which it, too, starts to burn.

A process which has a special appeal to those concerned with conservation is the conversion of solid wastes (which are predominantly organic in substance) into a compost which can be used as a combined soil improver and fertilizer. The process has made less headway in the United Kingdom than its advocates foresaw, and several local authorities who ventured into the compost business have withdrawn from it. Success in composting is not to be expected unless measures are taken to ensure that the raw materials contain no inhibitory substances in significant quantity. This is an unremitting task calling for the most assiduous application and control. The mechanics of composting consist basically of wetting the wastes (if they are not wet enough already), allowing the bacterial activity to proceed naturally, and turning the mixture from time to time. In its simplest form the mixing can take place in heaps on a suitable floor; in a more advanced technology, the whole process is mechanized, the wastes being fed into a slowly rotating horizontal cylinder which imparts a tumbling action as the wastes travel from the inlet to the outlet. The temperature of the mixture rises to around 90°C, at which level pathogenic and deleterious matter is said to be inactivated. An extension of the principle of compost making is to use sewage sludge instead of water for wetting the wastes. The advocates of this practice claim that the beneficial properties of the sewage sludge are imparted to the compost; the opponents point to the increased risk of inhibitory substances such as lead, copper and zinc being introduced by the sludge. In any event the process is not a balanced one, in that more sewage sludge is produced from a community than is needed to add to the refuse from that area.

The use of refuse as a fuel in steam-raising plants is attracting attention in several parts of the world. As the calorific value of refuse is, in industrialized countries, about one-third that of coal, it might be inferred that parity with the use of coal would be ensured if three times as much refuse could be burned. Moreover, as refuse is a low-cost (possibly even a no-cost) fuel, the balance, it might be thought, would be strongly weighted in favour of it being used in place of coal. There are, however, circumstances which militate against such a substitution.

For a start, the refuse cannot be burned in the state in which it is collected. It must first be prepared for burning by having most of the metal removed from it and by being pulverized, hammered, rasped, shredded or in some other way size-reduced to a 75 mm greatest dimension. The machines which bring this about are high in capital cost, and their operation makes heavy demands on labour, electrical energy and repair facilities.

Another detraction is the difficulty of matching supply and demand, for refuse is not a material that can be readily stockpiled. Another is the provision that must be made for the handling of a disproportionately large amount of ash. Another is the frequency of machine outage for inspection, cleaning and repair. Not only does this interrupt the generation of steam, but it necessitates having what may be quite elaborate stand-by facilities for the removal and disposal of the refuse that would otherwise have been burned.

It might be thought that, notwithstanding the shortcomings of refuse-derived fuel, an ideal application for it would be in the generation of electricity. Here, a difficulty of another genre arises. In Britain, if electricity is generated by anyone other than a generating or supply authority, the producer of the electricity may sell only to the authority any that is surplus to his own requirements. Payment is at rates comparable to the cost of generation by the authority's highly efficient plant and fuel, hence the financial inducement to produce electricity from the burning of refuse has been less than attractive.

Whichever way they are looked at, the benefits of burning refuse as a fuel are, at the best, unrewarding.

The management of the solid wastes from the extractive and metallurgical industries is confined, for the most part, to disposing of the wastes by dumping on land or in the sea. With the exception of slatey wastes and the red mud from the processing of bauxite (for neither of which there is much demand) the wastes can often be put to some profitable use for infilling low areas or for constructing embankments of roads. The best of the wastes may even be converted into materials of construction (cement, bricks, aggregates for concrete, blockwork, etc.) but this is rather chancy by reason of the impurities in the wastes.

Forecast and philosophy

Wastes are wastes for one of two reasons: because there is no known use to which the wastes can be put, or because there is insufficient incentive for anything useful to be done with them. When times are good, wastes are something to be got rid of as effectively and with as little offence as possible; when times are bad, wastes are looked at with renewed interest. Do they contain any hidden source of wealth that can be tapped? Can the cost of their disposal be reduced? Can the restrictions affecting their disposal be made less stringent without incurring unacceptable risks?

In the Far East, it is said, discarded condoms are much sought after, not with a view to their being reused as such, but because the rubber rings have a usefulness in the structuring of women's hairstyles. In Taiwan, it is said, pig dung is so valuable a commodity that its ownership is vested in the State which, because the theft of pig dung is on the ascendancy, is suffering economic loss. It is inconceivable that either of these practices could find an application in the United Kingdom, but the difference is perhaps one of degree rather than of principle.

Although there has been little that merited the term "scientific breakthrough" in the last sixty years, technological forecasting is as fraught with uncertainties in the field of wastes management as in any other. In the past the criteria that were employed in the selection of a waste disposal facility were cost and functional effectiveness. Cost necessarily took account of the amount of energy that was consumed in, or generated by, the disposal process, but energy by itself was not a commonly used yardstick. It is foreseeable that the Government might introduce sufficient incentive to ensure that energy took its appointed place in the scale of values alongside functional effectiveness and cost.

One likely development in the field of energy relates to the heat that is wasted because of the difficulty of utilizing it. This difficulty may arise from the absence of any obvious use to which the heat may be put, or from the fact that the heat exists at a low level. Such heat can, however, be used for heating water, for hot water is easier than steam to move over quite long distances, and at its destination it can be used for space heating or industrial process work.

Economic incentives or legislation may be looked for that would encourage or make mandatory the reclamation and recycling of certain components of wastes. In Sweden, for example, permissive powers are in operation to allow newly made paper to contain a proportion of salvaged paper and, in the future, the practice is to become mandatory. Paper which contains re-cycled paper is of poorer quality than paper which is made from virgin materials, because of the repeated shortening of the fibres; consequently there are those who are opposed to any scheme for the compulsory re-use of salvaged paper. Part of their contention is that trees, which are the source of the virgin paper-making material, are a legitimate crop, to be husbanded as one would husband grain or potatoes or grass. In the United Kingdom, paper that is to be in contact with food for human consumption must be of virgin material. But why, some people ask, do wiping papers and tissues have to be of the same immaculate quality?

Incentives might be given to encourage the processing of ores, etc., whose useful content is too low, according to present values, to make their processing economically justifiable. The same procedure might make it profitable for bings of colliery and coal-washery wastes to be gone over in the search for usable constituents.

An extension of any scheme of economic incentives or mandatory sanctions could affect the use of scarce materials in situations where their use is not imperative. The use of chromium for the plating of motor vehicle bumpers is a case in point. Another form of the same extension might be the enforced limitation of accelerated obsolescence which has the effect of bringing about the discarding of an article earlier in its lifetime than would otherwise be necessary. Organizations who undertake expensive research and development will have misgivings about the justice of any such move, for these organizations tend to regard accelerated obsolescence as a means of ensuring a just reward for their initiative and ingenuity.

The British Government's Department of Industry have promoted a scheme for the exchange of wastes between industries, the hope being that one company's wastes will be another's raw materials. Wastes from the processing of photographs and X-ray film might, for example, be taken over by firms who would recover the silver.

Solid wastes that contain organic matter in substantial quantity—domestic refuse being an example—provide more scope for technological innovation. First, if the means of separation could be improved, this would enable the last of the picking belts to be done away with and would be especially beneficial in relation to metals (aluminium, brass, tin, cast iron) which cannot be removed by magnetic means. Classification by air currents of different velocity or by liquid media of different densities seems to be indicated. So also does trajectory separation which takes advantage of the longer or shorter flights which wastes of different densities make through the air when they are given an initial impulse.

When refuse of this kind is cooked at a temperature of about 700°C in a retort sealed from the air, oil and gas are given off and a burnable "char" is left in the retort. This process is pyrolysis; it has been used for the treatment of house refuse but so far with indifferent success. It is, however, the best method available of obtaining some useful substances from discarded rubber tyres, and it is to be expected that further development work will widen the usefulness of the process.

It ought not to be assumed that it is only the solid wastes that are produced on a massive scale whose disposal is being studied in this way. Work is proceeding on a study of the usefulness of dried poultry droppings to supplement feeding stuffs for calves and other ruminants. Likewise feathers, hooves, and bristles are a source of protein, and the use of corn straw for the making of paper—long regarded as impracticable—has been given fresh impetus by the application of a new process based on the use of oxygen.

In the field of liquid wastes, one innovation has been the application of ultrasound (around 20 kHz) for the sterilization of sewage in ships. Improvement, however, rather than innovation, would seem to hold out the best hope for the future. If detoxification can be excluded on the grounds that it is not of general application, improvement may be considered under three heads:

(i) solid-liquid separation
(ii) biological processes
(iii) recovery and re-use

The methods of effecting solid-liquid separation in industry are for the most part gravity separation (sedimentation and flotation), vacuum filtration, filter pressing, centrifuging, and straining through woven wire fabric. Improvement is likely to take place along two directions: more efficient machinery and more effective pre-conditioning of the waste by additives. Higher mechanical efficiency will entail more resistant materials of construction, more efficient prime movers, and simpler methods of operation and maintenance. The nature of the additives and their application to solid-liquid separation allows more scope for research. Already, lime, ferrous and aluminium salts and polyelectrolytes have proved their usefulness, and it would be arrogant to pretend that no others are awaiting discovery.

Methods that are in use for solid-liquid separation (but not on the scale corresponding to that under discussion) are reverse osmosis, ion exchange, dialysis, evaporation, distillation and freezing. To them there might be added sonar vibration and magnetic field. All are worthy of further study, although the present indications are that something better than a mere physical scaling-up will have to be achieved if these devices are to play an effective part in wastes management.

Biological oxidation of organic wastes would seem to respond to the culture of more specific mirco-organisms. This has already been tried with varying degrees of success. As might be expected, the most successful application has been where the wastes are the least heterogenous in composition, and this points to wastes from a manufacturing process rather than from a community. An example of the use of specific bacteria is to be seen in the treatment of wastes containing phenols.

An innovation which is undergoing proving trials in the United Kingdom, although already well established in the United States, is to use oxygen instead of air, in the process of biological oxidation using dispersed micro-organisms. Two advantages which are claimed for the process are that it takes place in closed vessels (conventionally, oxidation by the use of air is in open-topped tanks) and the products of oxidation are more amenable to further treatment. In small plants "tonnage" oxygen might be used, but it is foreseen that for plants other than the smallest, the oxygen would be manufactured on site by the liquefaction of air. This raises the question: Would it not be worth utilizing the nitrogen which would otherwise be released back into the atmosphere?

One likely development in the treatment of liquid wastes is that in suitable situations the effluent will be used as a low-grade water rather than be returned to a watercourse. In the state in which it is produced, such water could be used for cooling, quenching and in the early stages of multi-stage washing. It seems unlikely, on grounds of safety or sentiment, that low-grade water will find ready acceptance in the home or in those industries where the operators would be in frequent or prolonged contact with the water. Nevertheless some see the use of low-grade water as the antidote to the wastefulness of flushing toilets with water of dietetic quality. The practice in some parts of Hong Kong might be adopted more widely. There, most premises are supplied with two, and some with three qualities of water—potable water for dietetic purposes, sea water for the flushing of toilets, and untreated fresh water for horticulture.

Envoy

A recent World Health Organization publication contains the statement: "Man cannot live in dignity in the midst of his own wastes". However many circumstances combine to make it necessary for man to live in such a state, lack of technological capability is not one of then. It, at least, is sufficient for the task.

CHAPTER FIVE

THE ARTIFICIAL ENVIRONMENT

JAMES K. FEIBLEMAN

Introduction

How often have we heard that "human nature never changes"? We might have stopped to consider that if this was the case it could be said of nothing else in the universe. But of course it is not the case, as the recently-acquired knowledge of pre-human types and early man has proved. In this chapter we shall look at the extent to which human nature has changed. The only way to encompass anything is to look at it from a sufficient distance. Here we shall do that for man, endeavouring to view him and his development as though he were a figure in a landscape.

It is generally acknowledged by those biologists who are concerned with the problem of human evolution that languages and material tools have played critical roles in the development of *Homo sapiens* from earlier hominids. The study of languages has been vigorously pursued by Wittgenstein and other in Great Britain, but the equally important bearing of material tools has been neglected. In what follows we must remember that we are not talking about ourselves, but about the species to which we belong. We shall be considering such questions as how that species got started and how it developed through the years. Briefly, the story of man is the story if how one animal learned to change his environment to suit himself, and how his reaction to that new environment changed him.

The emergence of mankind

Man is the product of his environment, and in a certain broad sense his environment is the universe—all of it. This must mean more than a billion galaxies, each with an average of a billion stars, and perhaps for each star eight satellite planets like those of our sun. If one out of every set of planets resembles the earth, then there are billions of occasions on which life could exist.

Everyone knows that the earth is a round ball of matter, but what is not so well known is that it is matter of an exceptional kind. Of the four states of matter: solid, liquid, gas, and plasma, only plasma is predominant. Plasma, which is said to be 95% of the material in the universe, is made up of atoms stripped of their electrons and therefore in an unstable excited state. It is of course uninhabitable, The other three have been described as "trace contaminants", although they are necessary conditions for organisms. Unbelievably complex, the world, then, is immense, and affords a richness of environment for humanity which is greatly in excess of what its equipment of awareness is capable of detecting.

The past, too, is immense. For humanity it was a tremendously large series of events which, however obscure, were certain to have a decisive effect upon the present and future; a long stretch of time, uncountable eons, during which it could be said of the human species that it did not exist and yet of the galaxies that they evolved and endured. A universe, then, of enormous size, occupying three dimensions of space having a large measure of symmetry, and one dimension of time which is asymmetrical with respect to the direction indicated by the succession of events. The direction of time is also that of the development of the complexity of matter. While energy in limited systems tends to decrease, matter in its living forms tends to increase in complexity.

The earth existed for four billion years before the appearance of life. The formation of amino acids may have resulted from electrical discharges in gases emitted by early volcanoes, occasioning an accidental development from inorganic chemicals. Slowly organic life arose, by chance, as it were; at first, only self-replicating compounds of nucleic acid and protein molecules; but then the Protista, the most primitive of unicellular organisms. All life had a common origin. Why physical events in planetary dust and gas should have led to biological organisms through anabolic processes is not well understood, except that in terms of large populations of instances the improbable becomes probable. Structures grew as instantaneous rates of function; and life, as strong concentrations of structure. The structures of life began to take up a kind of cumulative storage function.

For as long as a billion years there were only algae and sponges. Several hundred million years were required to move from fishes to amphibians, to reptiles, to birds and finally to mammals, by means of a biology based on a particular chemical reaction involving the synthesis by green plants of organic compounds from carbon dioxide and water, with sunlight as energy; and a reverse process involving the absorption of oxygen and the emission of carbon dioxide by animals.

Plants are stationary, and have a strong grip on the earth, e.g. the roots of alfalfa plants have been known to penetrate 8 m (25 ft) or more below the surface. But animals, especially man, are mobile and restless. The animal, man, derives his nature from other animals, from plants, and ultimately from rock erosion and organic debris.

There is one conclusion which can be reached from the foregoing account, and it is this. Since man as an animal emerged from his environment, as a product of it, there can be no reliable knowledge of anything in man than was not first in the environment. The implications of this proposition have been seriously neglected, especially in epistemology, but also in studies of man generally. Many of the new discoveries in mathematics, for instance, must always have existed as possibilities, even when there were no brains to know them.

Although some doubt still exists as to the nature of human origins, evidence seems to be accumulating that the fossil apes, *Dryopithecus* and *Sivapithecus*, living in the remote Miocene some 25 million years ago, may have been the true hominid ancestors. They were not highly-adaptive tree dwellers, lacking as they did the long powerful arms and legs of the great apes, and they shared characteristics in the upper dentition and maxilla with the Pleistocene tool-using species.

From the earliest specimens that could be described as hominids, it is clear that man has always depended on tools for survival. Culture began in fact with the first ape that grew weary of climbing. It has been suggested that perhaps the cold from an advancing ice age killed the trees and forced their dwellers to the ground. Somewhere between 1 750 000 and 600 000 years ago, the family *Hominidae* branched off from the family *Pongidae*. The great apes of Africa, the chimpanzee and the gorilla, developed in terms of arboreal life; hence the brachiating arms and fructiferous diet. The hominids developed in terms of terrestrial life; hence the upright posture, the use of hands, material tools, speech, and carnivorous diet. Man is inseparable from his ecological community, and this community is his material culture. However primitive such culture was at the start, it still gave him mastery over the other animals, many of which were more powerful.

In July 1963, in Australia, a rancher on horseback followed by his dog came across a herd of kangaroo feeding. The dog frightened the kangaroo and they fled, but the dominant male came over and pulled the man off his horse. They were struggling together on the ground, with the fight going against the man, when they rolled into a small stream filled with large stones. The man picked up a stone and with it crushed the skull of the kangaroo; the fight was over. Thus in 1963 the drama of man's device for winning the battle over naturally superior forces in his environment was repeated in the same primitive form in which it no doubt had occurred for the first time between a million and a half-million years ago.

There is nothing particularly new about the association of man with tools. They are evidently as old as man himself. The Australopithecine ancestors of man some million and a half years ago used stones and clubs as weapons. The earliest hominid ancestors of man lived in trees and required no artificial aids. He could sleep anywhere, and he could live on berries and fruit. Such arboreal life presented little danger from larger animals and few problems of survival.

Once on the ground and in the broad savanna, however, the situation became considerably different. The only food was the flesh of other animals, but the animals ran away and, in the attempt to catch them, a greater distance grew between the hominid and his prey. and so there were new problems.

He met them when he discovered that his power over his immediate environment could be greatly extended through the use of tools and speech. The early examples of both were crude. We have the testimony of stone tools and, although the speech has been lost, there were behavioural accommodations of a permanent sort. Bipedalism resulted from having to stand erect to run and throw stones. The development of larynx, vocal chords, tongue, and lips, resulted from repeated communication over long distances.

The humanization of the man-ape probably consisted in the process of externalization: learning to do outside the body what had formerly been done inside it. This constituted a distinct advance, because it was more efficient. The efficiency can be measured in two dimensions: intensive and extensive. The intensive dimension can be exemplified by cooking, which is a kind of external predigestion. The extensive dimension can be exemplified by writing as external speech, and by the use of stones as weapons to replace the similar use of teeth.

Within the Pleistocene, lasting from a million years and ending some ten thousand years ago, there were four ice ages. There were also a number of stages in the evolution of the hominids. Certainly they were intermediate types, and the development which led to them, like that which led from them, must have been a gradual process extending over a million years. The hominids travelled in small family groups and they were carnivores, especially the *Pithecanthropines*, who had body skeletons similar to that of modern man.

If we had more examples of hominids, we should see that the definition of *Homo sapiens* is not possible. The human species is not an absolute type, but rather an arbitrary division in what must have been continuous sets of organisms with almost imperceptible changes from set to set. The *Pithecanthropines* were more mobile and lived in larger social groups, and so made possible a greater gene flow. The diet of meat, instead of merely fruits and vegetables, freed them from the necessity common to so many animals for continual eating, while sight in open country took precedence over smell. The greater use of hand tools facilitated the development of speech, for the one seems to have stimulated the other.

The third and final stage before the advent of modern man was that of the *Neanderthalers,* who lived from 200 000 to about 40 000 years ago, and then disappeared. Although possessing a cranial capacity similar to ours, they showed marked differences, such as a recessive chin, prognathous jaws and large supraorbital ridges. They had progressed to such developments as core and blade tools, pitfalls, fire-making and shelters. They practised deliberate burial and ritual cannibalism; the latter they shared with other early types.

The tools had by this time become more sophisticated: flaked tools, spears and the use of fire, but a long time was required for their improvement. The hand-axe, which was used in South Africa and also as far away as England and India, remained the most efficient tool for 250 000 years. The use of tools brought abóut a strong selection pressure for an increase in cranial capacity. It rose in less than a million years during the Middle Peistocene from about 1000 to 1400 cm^3.

The emergence of society

There were other new developments. We have noted that when the early hominids took to the open spaces and adopted an upright posture and bidepal locomotion, long-range communication became imperative. The progression is an easy one to trace from this point. Stone tools and designated sounds were devised separately, but then came the next stage, with its marks, which are tools inscribed to indicate sounds.

The development of speech was no doubt responsible for the development of society. Internally as thought and externally as communication, individuals were tied together in the first formal organization. Languages furnished the alphabet of coded instructions on how to proceed to the next stage. Thought is a bodily skill conducted by means of signs and provides a more efficient form of retention. The brain of the full adult contains stored assumptions as to what the environment is like, which enables it to solve the problem of what he should do under exigent circumstances to preserve his existence. To this end the information has to be stacked in some kind of deductive order, so that it can be inductively available for application.

Most authorities today agree that the two factors which are most responsible for human development were the discovery of material tools and of speech, and for some authorities the material tools take precedence over the speech. As opposed to them, however, the modern school of structuralists in France, including men like Roland Barthes and Lévi-Strauss, insist that artifacts are signs, and hence susceptible to interpretation by linguistics.

It seems truer to say, however, that languages are artifacts. There are many activities other than speech, even though some of them involve speech. Which has historical priority is unknown, and both play important roles, but it would appear that for the use of tools speech was necessary.

Tools and speech may be grouped together under the term *artifacts*, and in this connection an artifact may be defined as *some segment of the material environment which has been altered through human agency in order to render it suitable for human uses.* Languages are always made up out of shaped sounds or marked surfaces.

It is possible to predict, at least to some extent, what man will do to artifacts, but it is impossible to predict what they will do to him. The journey from the pre-human types, recognized in pre-history, to the present stage of humanity was made in terms of the development of artifacts, i.e. in terms of their increase in number and complexity. The people we call *primitive* are those with few or no artifacts. There are many examples, e.g. among the most recently discovered are the Kreen-Akarores and the Txukahameis of Brazil, who are still living under conditions which prevailed in the Stone Age. They had never seen or used metal before the tribes were discovered in 1974, when a road was built through the jungle.

Once tools and languages had been invented, early man was perforce obliged to meet their requirements in order to continue using them. One of these requirements was adaptability; genetically determined behaviour was supplemented by learning and conditioning. Thus in the infant there is a readiness to accept information on authority, and later to adopt socially endorsed rules, such as moral codes.

It is at this point perhaps that society interjects itself as a new element in the situation, taking precedence over both the individual and the loosely associated group. The tools of communication have become complex enough to make possible both social organizations and transmission across the generations. Within the individual there begins to take shape the sets of shared beliefs and acquired skills which integrate him with his fellows in a system of behaviour patterns transcending his separate existence. His survival now in a certain sense is involved with the survival of others.

The first requirement of living is—life; the living organism seeks its own continuance. In terms of the human animal this means that his basic need is to survive, and for early man survival could be achieved only by hunting. Nomadic existence lasted a million years and, of course, was based on hunting. Such prolonged and intense conditioning was bound to leave its mark and we see it still in every kind of competition, including war, as well as in the more primitive love of hunting for sport.

Man in his present constitution goes back no more than 40 000 years. The earliest remains of *Homo sapiens* have been found in Borneo and date to that period. On this all the present authorities agree, and it is perhaps the most astonishing of all facts. For the first 30 000 years of this period he behaved very much like the previous species from which he must have mutated, i.e. he was a hunter, living parasitically on the herds which he followed through the migrating seasons. Killing must have been selectively bred into him, and only the best hunters survived. He fed and protected the members of his immediate family, but he fought with the leaders of other families with whom he competed for the spoils of the hunt.

Individual man, then, as a member of the species *Homo sapiens*, is not very old, and modern metropolitan man, man, the culture product, is very young indeed. It may be only a coincidence that the last ice age ended only 11 000 years ago and that the earliest civilization was begun a thousand years later. Agriculture is not over 10 000 years old, and industrialism only a few hundred. If, as we have noted, the greatest step in human history was taken when the earliest hominid came down out of the trees and subjected himself to the conditions prevailing in the savanna, the step from hunting to agriculture was almost as epochal. From the nomadic life of the hunter to the settled life of the farmer involved a change of conditions which altered the whole organism. The transition brought about a radical psychological shift: from an emphasis on feeling and the immediacy of the present, to one on reasoning and planning for the future.

The culture heroes who discovered the use of fire and invented the wheel had a companion whose work may have been even more important than theirs; for one day a genius arose who saw that the herds could be fenced in and the animals bred and consumed at will, provided also that the crops to feed them were cultivated. This was the beginning of animal husbandry and agriculture, but it was also the beginning of much more. It was in fact the beginning of civilization as we have come to know it; for now more-permanent houses could be built, many other artifacts produced, and the entire collection passed on through children to successive generations. There is evidence in Israel and Jordan of the transition from hunting to agriculture about 8000 B.C. The first wholly urban civilization took place in Sumeria about that time, and the first writing was invented there about a thousand years later. This kind of settled life led to all of the arts of civilization, and hence to modern man as we know him.

There have been no radical changes in man in the 40 000 years which brought him to his present condition, i.e. no known changes in his biological characteristics. However, it must be remembered that the phenotype is a result of the interaction of the genotype with the environment, and there has been a change in the environment. Within the last 10 000 years, man has invented an environment which is almost altogether artificial, with the consequence that he now lives in a world of his own devising.

By means of material culture, specifically through the proliferation and organization of specialized tools, man has made for himself a hospitable niche; and by then adapting to this artificial environment, he has improved himself biologically, at least up to a point. If that point was reached some 30 000 years ago, and no biological improvement has occurred since that date, it may be because biological improvement has been supplanted by improvement in material culture.

The change is so decisive, in fact, that it is too early to say how drastic its effects will be; but since it is taking man with it, the question for him is crucial. The rate of acceleration of material culture is already very steep, and increasing rapidly. Human development has reached a point at which progress is possible only in an artificial environment where artifacts constitute the "raw" materials needed in order to construct the next stage.

One way in which man differs from the other animals is that he has not only their interest in drive-reduction but also an added interest in terminal goal-achievement. The relation of man to his immediate environment is a sensitivity-reactivity system of energy-levels, with responses reverberating at every level. The immediate environment, containing as it does both artifacts and other individuals, consists in the same set of organizational levels, though it is customary to call them *integrative levels* when referring to their existence in the environment. The complete development of social man is possible only in an environment which is physical, chemical, biological and cultural, a total environment in which his goals can be attained and his needs reduced.

Genetic inheritance can transmit only from parents to children, whereas the epigenetic inheritance of material culture can transmit to an entire generation—obviously a great gain in efficiency. Lamarck was wrong, of course, about genetic inheritance, but there is a Lamarckism of the environment. There most certainly is an inheritance of acquired artifacts. It is what makes man civilized, an inhabitant of cities. Although tremendous anciently-transmitted forces come together when gametes meet to form the zygote, external development has substituted for internal development, until now skills and ideas can be combined in programmed machines. What man passes on to his progeny in addition to his genetic inheritance is a complex stress-producing material culture. Its course is totally known, but the individuals who can stand up to it will live to reproduce.

Externally, his capacity for invention has run the same kind of unknown course. The progress in artifacts was slow at first and was partly determined by the chance discovery of the appropriate raw materials. The Stone Age lasted a long while, but then the process accelerated. The Stone Age was succeeded by the Age of Bronze, and then by the Iron Age. Delicate precision instruments and massive earth-moving machinery were not even conceived until the forging of steel was discovered. Finally, with the finding of the light but strong tensile metals, such as aluminium and magnesium, there came a quick progress in the techniques of tool-making. Each new material brought with it new possibilities. The isolation and working of a new material has always disclosed potentialities undreamed of before.

The intellectual development of the human species—*Homo sapiens*—is due to the influence of material culture. After man invented artifacts, he was compelled to modify his behaviour in order to manipulate them; and it is the integrated collection of artifacts of many kinds, together with differential social groups and their established institutions, to which we have given the name *civilization*.

The emergence of civilization

The more complex the artifacts, the closer together men are able to live. Civilization is almost entirely a business of dealing with artifacts, of people in great numbers, packed tightly together in more or less permanent dwellings, guided by laws they have agreed to obey, speaking a common language they have recorded, and following customs they have established.

If the example of the other animals means anything in human terms, then crowding into a small space human beings who have become accustomed to occupying a much larger one, increases the intra-species tendency to aggression. For humanity, in other words, war becomes part of a way of life, as indeed it has been. In recorded history, which goes back securely for only a few millennia, there have been few years when there was not a war somewhere.

If we define aggression as the forceful alteration of some portion of the environment—conquering men or moulding materials (and the men as often as the materials)—then human life consists in some form of aggression. Among us there are those who exist only to fight, but for all of us it is always necessary to be ready to fight in order to protect our existence. Aggression in this sense may be constructive, or it may be destructive; it is a muscular need. Constructive aggression accounts for civilizations; destructive aggression accounts for wars, and we have had as much of the one as the other. Unfortunately, destructive aggression is much more efficient in reducing the need. It takes centuries to build a city, but only hours to burn one down.

Consider the forces at work within the citizen of any modern state. He is an aggressive animal who must dominate his total environment in order to reduce all of his needs. The needs themselves are organ-specific: water for the tissues, food for the stomach, sex for the reproductive organs, information for the brain, activity for the musculature, and security for the skin. The first three are obvious, but the last three call for some explanation.

It is well known that if the brain of the child does not receive a certain amount of information it will not develop. The lack cannot be made up in maturity. Thus information—knowledge—(and it can be false knowledge as well as true) is a crucial organic need.

The sedentary lives required for many civilized tasks make artificial exercise necessary. Artifacts have reduced the use of the muscles, which nevertheless make their demands. It may be the frustration of this need which breaks out periodically in wars.

The skin is the largest organ in the body, and it protects the body; to injure the body means in most cases to injure the skin first, and so it is the first line of defence. How shall the body survive? Somehow by linking the skin with the greatest forces in nature, either the cosmic universe or its cause; and, since this cannot be done literally, it is done symbolically, e.g. by contagious magic, by the Roman Catholic sacraments, by the laying on of hands.

Artifacts of special sorts are required for all of these need-reductions. Together with them man has produced an organization to house the collection: systems of political economy make possible the satisfaction of individual needs with a minimum of conflict.

Civilization involves a high degree of intensification, and hence the discovery of how to make aggression more effective; machines to alter the raw materials, and abstractions to concentrate the intentions; power-driven machines and absolute truths. It is worth noting that the scientific industrial culture thus far has consisted in the operation of highly complex machines and the carrying out of established systems of ideas which together are highly incongruent. Both have been inadvertently responsible for increasing the amount of violence.

The most alarming fact about human history is that in all of the 40 000 years since man has been in his present stage of development, there has been no progress in motivation. His aggression is ambivalent, he wishes to help and hurt his fellows, to aid the members of his in-group, and to assault those in his out-group. Man is half-disposed to be the friend of his own species and half-disposed to be its enemy. He is equally efficient at killing bodies and at saving souls.

With the sudden increase in population both the in-group and the out-group have grown enormously. In place of the tribe or clan, the in-group is now a nation or even a set of nations, and so is the out-group. The composition of both groups is fluid, e.g. in world war I Italy was part of the in-group and an ally, whereas in world war II it had shifted to the out-group and become an enemy.

The paradox in man's development is the constancy of his opposite aims. The only difference is in the efficiency and complexity of the artifacts by means of which he works toward them. Here we see almost in our own times enormous changes at the last evolutionary second. In primitive cultures the medicine man with his magic helped the in-group, while the warrior with his bow and arrow fought the out-group; now the medicine man has been replaced by the physician with his pharmacopoeia and his hospital, while the warrior has been replaced by the soldier with his inter-continental ballistic missile. Another example: help for the under-developed nations currently runs into billions; at the same time the practice of genocide continues. Witness the slaughter of the Jews by the Nazis in the 1930s, and more recently of the Ibos by the Nigerians in the 1960s.

Side by side with the paradox of ambivalent motivation there was one large progressive development, and it constituted a giant step forward. The externalization of organic function which, we saw in the last section, probably accounted for the emergence of man, now in its more complex form is responible for the development of civilization, e.g. libraries retain more than memory, reproductive devices can make permanent records of speech and even of behaviour, radio extends the human voice to the entire world, computers calculate faster than brains. The automobile and the airplane have greatly extended transportation beyond the ability of anyone's legs. The bulldozer and the crane may be regarded as prosthetic devices, extensions of the musculature and, indeed, since they can be operated by means of levers and push buttons, almost substitutes for them. Most important of all from the social and cultural point of view, all of these extensions of human organs can be employed by more than the members of a single generation, they can be handed down intact, often for centuries.

The acceleration in the development of material culture has been as thoroughgoing as it was sudden. The use of artifacts on a large scale is immensely effective. It was what Portmann has aptly called man's "artificial second nature". By now there is not much in the available environment which man has not greatly altered, including the ground he walks on and the air he breathes. He has cut down the forest, built cities, farms and grazing lands in the clearings, brought oil and minerals to the surface, and filled the air with smoke and dust. The oceans are perhaps the least affected by human activity, but even they are not entirely free. Surface travel, submarine exploration, and the dumping of waste products are some of the ways in which the ocean, too, is affected. Thus man lives in a world of his own making, largely on terms which he himself has devised.

The interaction between man and his environment changes both, but the process affects man more than it does his environment; for only the immediate part of the environment is changed, whereas the whole of man is. His entire background consists in one vast collection of artifacts. The future effects on man of his epigenetic inheritance are sure to be as large and dramatic as the effects of his genetic inheritance have been in the past.

Homo sapiens is a polytypic species, and all men belong to it in virtue of the existence of a set of intercommunicating gene pools having biological continuity; social barriers in no wise constitute isolating mechanisms. No racial difference has been established for even a single mental trait. Cultures are the result of diverse responses to varying conditions; and, although cultural differences are profound, they are not biological. All men have the same needs, but not all live in the same environment; and methods of need-reduction are always designed to accord with the environment. Judged by the similarity of phenotypes, it would seem that cultures operate to impose uniformities on individuals, despite the wide diversity of genetic variation.

Culture may be defined as an organization of men and artifacts interacting both with each other and with the non-human natural environment. Cultures operate by improving technologies, working over of materials in order to bring out their desirable potentialities, and making certain of the possibilities of materials by transforming them. What can be done, employing ink marks on paper as directions for how to use an old construction of glued wood and cat gut, is illustrated by a Mozart concert, with a Menuhin playing a Stradivarius violin. The individuals who are capable of such a performance are any whose talents and training have been requisite, irrespective of biological origins.

The building of a specialized environment by industrial man has led some authorities to suppose that he has been set free in such a way that his own development is unimpeded. The truth would appear to be quite the opposite, for man is more tightly integrated into his environment than ever before. Early man might have been able to survive in any non-human environment on the surface of the earth, but modern man requires the culture he has constructed, and could not survive very long without it. He has bound himself into a particular kind of ecosystem.

Thus, far from being independent of his special environment, he is more than ever dependent upon it. Men in advanced industrial cultures are in danger of existing only to tend machines. Thus tools may get turned around and, instead of being employed by men, may instead employ them. They threaten to dominate their inventors and users by assuming too large a role, functioning as ends when they were designed as means. The automobile, for example, was intended to serve the need for rapid and efficient transportation. It has, however, accumulated other functions: it may be regarded as a beautiful object, treated as a status symbol, or employed as a deadly weapon. The amount of time and energy devoted to the maintenance of machinery may be out of proportion to its usefulness.

What is true of material tools may be true also of languages. Signs of an arbitrary nature are developed not only as names for single material objects but also for classes of such objects; and such class names are combined in sentences, and the sentences themselves combined into entire systems of ideas. In this way abstract systems are brought into existence.

Now it is possible for men to employ abstract systems to their own purposes, chiefly as the explanatory schemes of philosophy, science or religion. In this way the need to know is effectively reduced. However, the same danger exists for ideas as for tools; they can come to dominate men instead of being dominated by them. When believed to be the absolute truth, they tend to control the believer. In short, men who begin by using systems of ideas may end by being used by them, and so become the victims of a system which exercises a tyranny over their thoughts, feelings and actions. Men in the grip of the absolute truth of systems of ideas will slaughter other men without the slightest compunction, as they have done for millennia, and as fascists, communists and religious fanatics generally are still doing in our own time. It is a sickening story from which no large group is immune. Systems of ideas in conflict tend to drag men after them to kill or be killed.

Modern man can exist in a capital city only so long as he and his fellows share a common set of beliefs. He has adapted to its requirements so completely that he is unprepared to meet any other kind. The urgent necessities of both cooperation and conflict set the conditions for him, and bring him with equal force to the brink of both love and war. What this will eventually do to him it is far too early to say. Evolution is a slow process even at its most rapid, and the newest development is too new to evaluate.

Due to the influence of gene flow and selection pressures, culture branches off, but biology does not. There are greater differences between individuals of a given culture than there are between races. The newly acquired capacity of man to objectify and externalize his ideas and skills has not only transformed the surface of the earth, but has also enabled him to have access to a detached view of himself. He has moved from a position of egotism to one of self-awareness, and from there to an objectification of the elements of experience. He has learned in the process not only that each individual is unique before birth, but also that both genetic inheritance and material environment contribute to his traits, a progress peculiarly hominid. His concerns have been transformed into world conditions.

The continuance of evolution

Let us pause at this point to see where the argument has brought us.

Man, however primitive, always lives in a world containing a few artifacts, but civilized man lives in a world composed largely of artifacts. The things with which man is chiefly concerned are those which he has altered to suit his own needs. Very little is as it was before he made his alterations in it; certainly not the earth on which he stands, nor the air he breathes. The earliest history of man is also that of his tools and signs; we have noted that he did not exist apart from them. Yet the fact that without man there would be no tools and signs has perhaps blinded us to the companion fact that without tools and signs there would be no man. Artifacts have enabled man to develop upright posture, unusual skills, brains, languages, roles, institutions, and indeed entire cultures.

Human individuals are voluntarily conditioned by artifacts. They will respond with chain-learned behaviour to the stimulus of the artifact. The voluntary part of this procedure is the extent to which the individuals submit themselves to such training. The method is one similar to that of the natural selection Darwin has made familar in organic evolution. Through a process of selective responses, the individual learns to repeat only those ways of operating the tool or sign which prove most effective, and not to repeat the others. Due to the peculiar construction of both kinds of artifacts they can be dealt with only in certain preferred ways. Thus they condition human behaviour. Their intractability is a brute fact with which human behaviour must reckon.

What then does this mean in terms of human evolution? All students of the way mankind has developed from primitive archetypes over the last million years have acknowledged the prevalence and persistence of artifacts.

In a sense culture represents the efforts of man at self-perpetuation—for he is culture-bound. He does not live in a world of his own. He lives instead in a culture his ancestors have made for him, one which he and his contemporaries have inherited and modified. Culture is the rearrangement by man of his available environment. He is certain to see his own development, not only through the advantages and limitations of that culture, but also through those of himself as an individual. Thus the effort of each man is primed by partial success, but also doomed to partial failure. It marks another gravestone bearing the same haunting memorial:

> Here lies the remains of one more man who wanted to be more than man by encompassing more of being than human being.

If man has a leading principle it is this: he is that animal which endeavours to exceed itself. Every mark he has left upon his genes in the successive stages of his evolution, every scratch upon every stone which has survived through the process of external inheritance, offers evidence in support of the truth of this principle. He does not act merely in the interests of reducing his needs, but reaches out beyond them to make contact with the rest of the cosmic universe through thought (e.g. the philosophical systems), through feeling (e.g. "supernatural" revelations), and through action (e.g. the exploration of space).

There is a sharp distinction in this regard between European and Asian man. European man has sought control over his environment, Asian man has sought control over himself. The former is outward bound, the latter inward bound. Both have pursued their goals in depth, which is to say, in excess. Both have sought to exceed themselves, the European in material construction, the Asian in psychological transformations.

Animal needs are human needs. Where the human departs from other animals is in an understanding that, with certain alterations due to his own overt behaviour, material objects in the immediate environment can be made to produce the goal-objects necessary for need reduction where none existed before. An empty field and seed-corn cannot of themselves satisfy hunger, but with the proper procedures they can be brought to do so. By means of instrumental investigations and mathematical calculations, inquisitiveness in certain directions can be satisfied. Understanding means acquiring techniques, the knowledge of formulas, and the possession of the requisite skills. Such equipment is externally inherited as part of the culture. It must be learned afresh by each generation of individuals. Such acquisition is called *formal education* and is usually specialized: languages, mechanical aptitudes, professions.

We have noted in the last section that the needs are animal needs, and that as basic tissue needs they are furnished with organs. The best interests of the individual are served by the reduction of his organ-specific needs, and the drives function to reduce them. But once a drive is fired, it acts on its own and cannot be recalled; thus the effects which it has may run counter to the corresponding need. Drives acting in this way are powerful and often blind; they do not necessarily stop when the needs which have occasioned their activities are reduced, but continue on, dragging the individual after them.

Needs are physiological, drives are psychological, and the goal-objects are material artifacts of cultures. Needs can exist without drives; indeed most of them must, since many needs coexist, whereas, ordinarily, overt behaviour can exhibit only one drive at a time. But drives cannot exist without needs; a drive is the activation of a need.

The material artifacts of cultures were brought into existence by the alteration of material objects which involved human action of some sort and could not have existed without it. The interactions between human individuals and material objects take place by means of overt behaviour. It results in alterations within the central nervous system of the individual (learning) and alterations of the material objects (the construction of artifacts). Such interaction is a continuing process; as the artifact is altered it affects the organism, and the cycle of interactions comes into existence. Thus we are confronted with the phenomenon of a brain-artifact circuit in which material cultures are produced and individuals made over into members of them. (It is worth noting in this connection that every artifact invented in imitation of the brain, such as computers, reduces further the belief in the difference between mind and body.)

The individual does not always act in his own best interests. For instance, the drive for sexual satisfaction may continue long after the sexual need has been reduced, becoming more like the generic need to dominate the environment in its Don Juan phase of collecting conquests. Again, the drive for survival may become so obsessive that it continues long after the individual has been reassured. Any drive may in this way pre-empt the individual's entire energy and become his whole reason for existence, thus inhibiting all other drives. This blocking effect produces the exaggerations of action which the ancient Greeks recognized so clearly as "outrageous behaviour", but it produces also the self-sacrificial devotion to causes, and it produces the monumental institutions of religion, science and art.

Whenever the environment has changed appreciably, the species has had to adapt or perish. There are cases where this challenge has been met successfully, e.g. by the horse, and cases where it has not been met, e.g. by the dodo. Where the environment has not changed quickly or appreciably, some organisms have survived unaltered for millions of years, as has the horseshoe crab. A change in the environment is responsible for new stimuli which confront genetic mutations with new selection pressures. In this way the environment stimulated the species to adapt and so to survive.

In man alone there has been a new and very different sort of adaptation; for man has gained sufficient control over his environment to enable him to decide just what his stimulations shall be. The artificial environment which he has produced by effecting his own modifications of the environment has won for him an external inheritance which in many ways is as important as the internal inheritance. The internal inheritance is the genotype, the totality of genetic factors; the external inheritance is material culture—the integrated system of tools and signs which are those parts of his environment by means of which he has been able to change the other parts.

It is possible that individual man's early association with his fellows was complete, and like the starlings today they wheeled and turned as one. There was a period when tools were world-wide, such as the hand-axe in the Lower Palaeolithic, and it has been supposed that perhaps the differentiation of tools was what first divided men into groups which both gave the individual a feeling of belonging to a group and at the same time generated a hostility to other groups. Our knowledge of prehistory is so nearly nonexistent as to render such ideas purely speculative.

The world of material culture in which man lives is as important to his development as his genetic inheritance, which turns out to be the inherited part of the environmental influence which has survived through adaptation. By means of material culture he has been able to effect changes in his own existence which he could not otherwise have done. His need for survival has been reduced through the extension of life expectancy in scientific-industrial countries, where it has almost doubled. In this high adventure the individual is inevitably fully immersed. The social and cultural milieu is more than any one individual could surmount or even encompass. He can only endeavour to participate in it to the best of his individual capacities and, perhaps, if he is that rare specimen, a productive and originative individual, influence it a little.

Philosophy in practice

There is a logical structure to the life of man. Where he begins is determined by the equipment he brings with him to his birth, and it is considerable. He inherits the past of his ancestors, and thus acquires all sorts of capabilities and limitations; but he acquires during infancy the responses to artifactual and social stimuli. He is in contact with tools from the cradle, and adults make signs to him in it. Both affect him dynamically because of his inheritance of that extreme form of adaptability called *learning*, which is the ability to elicit from responses the capacity to respond. He is also to some extent unique, for there is always a genetic variation peculiar to each and every individual. There is much in this equipment which is disparate, but through combination in a single organism it acquires a unity.

The structure of the human individual is that of a living and acting organization, and its life cycle has a form which is akin to that of any logical system laid down along a time-line. The early events in the life of the human individual have the character of axioms. Whatever the infant and young child does will determine the theorems which are represented by his behaviour in later life. The rules in inference are artifactual as well as social; they determine the limits within which activity is possible, and even to some extent how it shall be determined.

Life is played out upon a large artificial framework of potentialities within which the individual, bearing with him his inherited character as well as his own early self-determinations, acts as an individual. The theorems of later life work out in practice the possibilities laid out for him in his earliest years by forces which lie only partly within his control. Among the factors are chance and its rules, the logic of his inheritance, and the influence of his personal history, partly determined and partly probable, aimed perhaps by his own resolution moving within material culture among the debris occasioned by the intersection of factual variables.

The existence of the individual as an integrated and dynamic organism presupposes a direction to human life. The aim of life is the perpetuation of life, through immediate survival and ultimate survival. Man's primary aim is survival, then, and the individual works for survival through his efforts to dominate the environment; hence his basic and characteristic aggressiveness. The direction of life consists in the reduction of the generic drive of aggression, the aggrandizement of the ego, the need of the individual to exceed himself and to extend himself into greater and greater portions of the material world. The aim of the reduction of the importunate needs is to gain precious time and energy for the pursuit of the reduction of the important needs; and the aim of the important needs both separately and collectively is the permanent continuance of existence. The individual does not wish to die, and his strategy is to so spend his life as to avoid it; or, failing that (as fail he must), to cheat death by employing life as a gateway to immortality through concrete achievements—a work of art, a scientific discovery, a church.

The life of the individual is a continual development. It is never static, and so the reactions are never twice the same. The needs are lacks in some organ or in the whole organism, but they are stimulated by artifacts. The activation is supplied by the brain-stem reticular formation, the hypothalamus, or the limbic system. The drives are neurohumoral, but the overt response is psychological. Behaviour is best accounted for by a generic internal inheritance so far as the biological factors are involved, but by material culture so far as the external inheritance is concerned.

In most of the important instances of man's needs, the hope of reduction lies in the alteration of some material object outside him. He became human in the first place by extending himself into artifacts; he can improve the human situation in the direction of maximal happiness and minimal pain only by further efforts in the same direction.

The individual who dies may still have contributed to the human succession in two ways. He may have been responsible for the birth of children, in which case he will have passed on to them his genetic inheritance, and thus served as a link from the past to the future. He may have contributed to the material culture in which he and the other members of his society have lived immersed, in which case he will have passed on to succeding generations his contribution to the external inheritance. The former way is more animal, the latter more human.

Had not a few individuals added to the material culture, neither it nor its passage through the years as an accumulation would have existed. In this the individual did at least exceed himself constructively. Like the tiny creatures which collectively build coral islands, he donated his energies in a way which ensures that culture shall endure, at least for a while. Beyond himself man makes a social contribution. It is possible and—so far as our limited knowledge goes—even probable that this is the principal part of the destiny there is to a man.

It should now be possible to state the human problem more succinctly.

Man, Nietzsche said, is a bridge from ape to superman. We have noted already how the span from ape to man was constructed, now we must guess how the span from man to superman could be projected. The individual must first become acquainted with the nature of the savage animal forces which still exist within him, and secondly he must learn how to redirect them into exclusively constructive channels. Socrates' command "Know yourself" is one which has not yet been obeyed. In the framework of modern biology it is clear that individual peculiarities are not what he meant, not the personality in so far as it differs from others, but the organ-specific drives which each individual possesses and all share.

Modern man is capable of greater efficiency than the ape. He can control his environment to such an extent that it has enabled him to master all of the other animals, to multiply exceedingly, and to prolong his own life span. Superman will have yet one more task, and it is as large as his previous ones. He will have to learn how to master himself. So far this is something he has not yet accomplished—for the accomplishment waits upon understanding. Before the superman who can control himself can be brought into existence, man must first comprehend the nature of the problem, he must first understand his own needs and capabilities. Understanding must precede control; he must know at first just what are the forces with which he is dealing, and then perhaps devise a method for directing them. It is unthinkable that such forces should be cancelled, for they are of the very nature of life; but they can be redirected into constructive channels.

On the subjective side, before there can be a superman, present-day man must learn how to develop his sensory receptors. He will need both a great sensitivity for the organs he now has, and greater reasoning powers to interpret their results. Only through inquiry can he hope to improve his knowledge and so extend his control over himself and his environment. Only through inquiry can he hope to learn how best to serve society.

The prospect assumes a stable society in which scientific progress continues unimpeded. The experience of scientists in the Soviet Union and in Red China would argue otherwise. When politicians are the ones who decide what science should investigate and what it should not, it is only a matter of time before progress in science is brought to a halt. Therefore the basic question may be formulated somewhat as follows.

Will science be permitted to continue until it can develop a science of society?

This is an important question to which at the moment no one has the answer. Meanwhile we await a solution, and it could conceivably come from either of two directions, from the physiologists through the control of motivation, or from the technologists through the control of the material culture. We know only that all of a sudden man has been delivered over to his own responsibility, and he has the capacity to redesign himself in two ways: by altering the environment, and hence himself indirectly, or by altering the genetic code, and hence himself directly. What should he want in his successors? This is a philosophical question.

It is easy enough to see what must be done, though not as easy to discover how to do it. The superman of the future will be one who could have developed his tools and his languages to the point where he is the complete master of an immediate environment which through his efforts has been greatly extended.

There has been an advance from the stereotyped behaviour which produced always the same level of primitive culture to the adaptive behaviour which has produced accelerating civilizations, or, in other words, a movement from mechanical responses to dynamic responses. Now there is a further step in sight, and it consists in preparing material culture so that specimens of it can conduct their own stereotyped and even adaptive behaviour without the need for a human operator. Examples are: the thermostat, the stabilizer, the computer—perhaps also the nuclear reactor.

Perhaps this kind of behaviour can be called *instigative behaviour*. Instigative behaviour looks to a complete act of externalization, with aim, mechanism and process together in one package, all independent of the human operator and going as far in the direction of complexity as levels of self-reproduction, as in the model of the Turing machine. This, no doubt, is the direction of the scientific-industrial culture, though only the smallest steps toward it have thus far been taken.

There are even further vistas. Man may learn how to grow with a language which in complexity and subtlety makes possible works of art which deepen all his experiences. He may intensify his equipment both for self-improvement and for self-destruction. Everything will depend upon his ability to organize all of existing human society so that wars are no longer possible, and upon his ability to find substitutes for his need for aggression, so that wars are no longer necessary. He will have controlled and planned the world population, he will have extended the ceiling over the human span of life, and he will have succeeded in bringing nearly everyone up to it.

Superman will know what he does not know. He will know, for instance, that he has little if any certain knowledge, that all his actions therefore must be taken on the basis of probabilities. A man with a religious faith, who still knew that certain knowledge was not his, would seek converts through reason. Actions follow from deeply held beliefs, i.e. from implicitly-held philosophies, and beliefs after all have the quality of feeling; but the feelings can neither be accepted by feeling nor defended by feeling, only by reason.

This is the Platonic method which the philosopher, A. N. Whitehead, once referred to as the victory of persuasion over force. It will know no victory until men have come to an understanding that any probability greater than half is quite sufficient to guide action, and that action taken on such a basis is apt to do little that could not be undone if it were proved wrong.

Thus superman will hold probable faiths and statistical beliefs, and he will engage in tentative actions. Remember that it is the survival of the species that Darwin was talking about. Natural selection operates to preserve species, and is wholly unrelated to the fate of any one individual.

Artifacts and human survival

The emphasis in this chapter has been on the species and its individual members— how they have emerged, multiplied and prospered, as a result of the discovery of artifacts. There is a further point which must at least be touched on before the discussion can be properly ended. It concerns the artifacts of that wider society we call civilizations.

The larger and more advanced the society, the more complex its artifacts. These are not thrown out by chance, but are well integrated with the social group into a workable system. And the system as a whole breaks down into those divisions of labour, involving men, skills, ideas and artifacts, which have become established as institutions, such as are found in transportation, communication, government, and the like.

Institutions differ with respect to the kind of assistance they provide—usually, however, including buildings, tools, and material symbols. Each has its own special variety of artifact. A single example should suffice in illustration. Look at the special artifacts of religion and the enormous importance they carry. Sacred relics, whether the thigh of St. Francis Xavier or the tooth of the Buddha, convey rather large and extended meanings, and it is useless to try to save the meaning from the relic; they belong together. The same can be said of the preservation of political charters, of ancient works of art, and of other such specially prized material objects.

Consider, for example, the means of transportation in an advanced culture: the motor-cars, trains and airplanes which are employed to meet the needs for social mobility. Similarly, consider the avenues of communication, the books, periodicals, postal system, wireless and television networks. Obviously these are not random artifacts but are well-integrated parts of a whole in which even the social groups are counted as components. Men and materials are tightly organized together, and neither could function without the other. Each item—man or machine—has its separate task to perform, but all work together as functioning units in a highly intricate organization.

The underlying scheme upon which the entire edifice rests and operates is the established charter of the society. The political system with its economic components maintains a certain established order which had been earlier agreed on. Behind that lies a silent system of ideas concerning what is primarily real, an implicit philosophy which is rarely acknowledged, and even more rarely brought to the surface, but one which is held nonetheless as a self-evident set of absolute truths.

Society, or civilization if you wish, is structured and moves as a whole. It is also in terms of these truths that the state is defended and more often than not expanded. It is these that clash when societies come into conflict; for, come into conflict they inevitably must.

Every culture has its own set of absolute truths, and for absolute truths men are willing to kill and even to die. That such systems of ideas are dangerous when established is only a measure of their usefulness in reverse; for, if they make the construction of a culture possible, they also make its destruction necessary. This is the pass that the discovery of artifacts eventually comes to: that there can be human cultures, and that human cultures can break out into open conflict. Artifacts furnish the means for living peacefully together in great numbers; they also furnish the highly efficient weapons of war.

It may be possible to have the one without the other, but no one has yet found out the way. Thanks to the extraordinary acceleration in the techniques of material construction, we are able to build nuclear weapons as well as apartment houses, streets and hospitals, with equal efficiency. And the one may spell the end of the other. It has happened that we have lost control of our artifacts. The alteration of the environment has meant also the pollution of the environment. A nuclear war would only mean a further pollution, but one that could possibly extinguish the endangered human species.

Thus the end of the story that began when the first man stood erect in order to throw a stone at a small animal nearby may be spelled out in his being able to fire an intercontinental ballistic missile at a large urban population continents away. He may be overwhelmed by the material civilization which now dominates him.

It is too late to turn back. The romantic desire of the idealists to "return to nature" by which they mean, of course, to the use of artifacts which did less to disturb the environment (for no one can live without artifacts of some sort, be it only a bow and arrow and a device for making a fire), cannot be fulfilled by millions of people, whether it works for a very few or not. Three hundred million Europeans, eight hundred million Chinese, cannot return to the woods. It takes all of the resources of the scientific-industrial culture in the middle western United States moving on the problem of agriculture to feed the Russians as well as the Americans.

Evolution works in the direction of greater order. The next stage in the development of the human species would probably be a global state. This would involve making a system of order which was world-wide. If it were to be like the present social orders, based on absolute truths, as for instance the communists base their societies on Marxism, it would not be workable, for absolute truths exist only in closed systems which do not admit the possibility of alternatives, and a global state would not be a closed system so long as other worlds are possible. If it is true that the amount of information in a system is a measure of its degree of order, it is also true that the enlargement of the system, which must be made irregularly, is also a function of the number of its degrees of freedom.

Thus, if it is too late to go back, at least it is not too late to look forward and to see that international cooperation is the only hope of survival. Helsinki may be Munich all over again; and the end-product of détente, of the American conciliation of Soviet aggression, may well be a nuclear holocaust, which every man in his right mind would do anything to avoid.

Pushed too far, however, any animal will turn and fight. We must learn to avoid that at all costs, but to control our artifacts means also to begin by controlling ourselves. There are forces within us, as we have noted, which make for aggression in its destructive phase and not merely for construction. We want to keep the fine arts and pure sciences by all means. But in ways that we are as yet at a loss to understand, doing away with the destructive tendency does away also with the constructive, as happens for instance in prefrontal lobotomy; and so the solution cannot lie in that direction.

Social control begins with individual control, the control by the individual of himself, and an ordering of the priorities of physiological needs. We are very far from understanding just how this can be done, but the solution to the problem is an urgent necessity.

FURTHER READING

Clark, G. (1969), *World Prehistory*: A New Outline, 2nd ed. (Cambridge University Press).
Feibleman, J. K. (1963), *Mankind Behaving*, (Springfield, Illinois, Charles C. Thomas).
Hawkes, J., and Woolley, L. (1963), *History of Mankind*, Vol. I, *Prehistory and the Beginnings of Civilization* (London, Allen and Unwin).
Hebb, D. O. (1949), *The Organization of Behavior*, (New York, Wiley).
Waddington, C. H. (1960), *The Ethical Animal*, (London, Allen and Unwin).
Young, J. Z. (1971), *An Introduction to the Study of Man*, (Oxford, Clarendon Press).

Index

DATE DUE